知りたいことがすぐわかる
数・式・記号の英語

鵜沼 仁 著

丸善出版

まえがき

　技術立国を目指すわが国としましては，今後ますます技術文章の翻訳が必要となります．工業高校・高専の先生や，大学・大学院の学生および先生，さらには各製造業に携わる技術者が研究テーマを国際舞台で発表する機会もますます増えていくものと思います．

　論文や研究テーマには数や，数式を含む文章が含まれますし，これらの数や数式を正しく英語で書き，しかもその内容を正しく聞き手に伝えなければなりません．

　例えば日本語で，「$V = \pi r^2 h$ を r に対して偏微分すれば $2\pi rh$ が得られます」という文章を口頭で伝える必要がある場合，偏微分をどのように訳し，しかも $V = \pi r^2 h$ や $2\pi rh$ をどのように読むかがわからなくては，結果的にこちらの考えが正しく相手に伝わりませんし，技術発表も成功したとはいえません．

　本書はこれらを考慮し，数や数式に重点をおいて下記にもとづいてまとめました．

・数や数式を含む技術文章ができるだけ容易に翻訳できるように，多くの例文を記載した．
・さらに数や数式を含む技術論文や研究テーマを英語で発表する場合を考え，数や数式などにはそれらの読み方を併記した．

　数や数式を含む文章の読み方については，もちろん，様々あります．ここに記載したものが唯一ではなく，一般的な理解しやすいものであることを申し添えます．

本書によって，複雑な数式も容易に読むことができるようになり，さらには本書に記載されていない数や数式への応用も容易になると思います．

　本書が，読者に多少でもお役に立てれば幸せです．

　　2003 年 10 月 30 日

　　　　　　　　　　　　　　　　　　　　　　　　鵜沼　仁

目　次

英語索引　xi
索　　引　xxvii

1 章 ● 数と数を表す数字と文字 ——————————— 1

数　1
数字　4

2 章 ● 数の集団 ——————————— 13

代表的な数の集団　13
そのほかの数の集団　17

3 章 ● 四則演算 ——————————— 19

四則演算　19
　構文例および読み方　22
指数，べき指数　26
累乗，べき　26
　構文例および読み方　29
対数　32
　構文例および読み方　34
根号　36
　構文例および読み方　38

分数　39
　　構文例および読み方　　41
小数　43
　　構文例および読み方　　47
四捨五入する，丸める　49
　　構文例および読み方　　50
比率，割合，〜の比　54
　　構文例および読み方　　54
比例，比例算　55
　　構文例および読み方　　56

4 章●線 ——58

直線　58
線分　58
半直線　59
交差線　59
垂線　59
平行線　60
横断線で切られる平行線　61
　　構文例および読み方　　63

5 章●統計 ——64

平均値，または，算術平均値　64
　　構文例および読み方　　64
重みつき平均値　65
　　構文例および読み方　　66
中央値　67
　　構文例および読み方　　67

最頻値　68
　　構文例および読み方　　68

6章●多角形 —————————————————————69

三角形　69
　　構文例および読み方　　71
四辺形　73
そのほかの多角形　76
　　構文例および読み方　　77

7章●円，扇形および楕円 —————————————79

円　79
扇形　81
　　構文例および読み方　　82
楕円　83

8章●立体図形 —————————————————————85

球　85
円柱　86
すい（錐）　87
多面体　88
角柱　89
角すい　90

9章●直交座標系 ————————————————————92

直交座標系　92
そのほかの座標系　94
　　構文例および読み方　　94

目　次

式　98
　　　構文例および読み方　99
　　多項式　100
　　　構文例および読み方　101
　　方程式，等式　102
　　　構文例および読み方　103
　　不等式　104
　　　構文例および読み方　104

10章●微分および積分 ─────────────────105

　　(1次) 微分　105
　　2次微分　105
　　n 次微分　106
　　　構文例および読み方　106
　　(1次) 偏微分　107
　　n 次偏微分　108
　　　構文例および読み方　109
　　積分　110
　　多重積分　111
　　　構文例および読み方　111

11章●三角関数 ─────────────────113

　　三角関数　113
　　　構文例および読み方　114
　　逆三角関数　115
　　　構文例および読み方　116
　　双曲線関数　116
　　　構文例および読み方　117

逆双曲線関数　117
　　　構文例および読み方　117

12章●級数 — 119

　等差数列または算術数列　119
　　　構文例および読み方　120
　等比級数　120
　　　構文例および読み方　121

13章●順列および組合せ — 123

　順列　123
　組合せ　124
　　　構文例および読み方　124

14章●和，合計または総和 — 127

　和，合計または総和　127
　　　構文例および読み方　128

15章●行列式および行列 — 129

　行列式　129
　　　構文例および読み方　130
　行列　130
　　　構文例および読み方　131

16章●そのほかの記号 — 132

　そのほかの記号　132
　　　使用例および読み方　133

英語索引

a to *b*	$a:b$ (a 対 b)	54
abscissa	横座標	93
absolute number	無名数	17
absolute value bars	| |	21, 132
	絶対値表示バー	21, 132
abundant number	過剰数	18
	豊数	18
acute angle	鋭角	63
acute triangle	鋭角三角形	71
addend	加数	19
addition	加算	19
additive inverse	加法に関する逆元	25
adjacent angle	隣接角	61
adjoined number	添加数	18
after a certain decimal place	ある小数点以下	49
algebraic irrational number	代数的無理数	16
algebraic number	代数的数	18
alternate angle	錯角	62
alternate exterior angle	錯外角	62
alternate interior angle	錯内角	62
altitude	高さ	69, 86, 90
angle	角度	61
	∠	61
antecedent	前項	54
antitrigonometric function	逆三角関数	115
approximate	近似する	49
approximately equal	ほぼ等しい	16
Arabic numerals	アラビア数字	1
arc	円弧	79

英語索引

xi

English	Japanese	Page
arccosine	アークコサイン	115
arcsine	アークサイン	115
arctangent	アークタンジェント	115
area of a circle	円の面積	80
area of a trapezoid	台形の面積	76
area of a triangle	三角形の面積	70
arithmetic mean	算術平均値	64
arithmetic progression	等差数列	119
arithmetic sequence	算術数列	119
array	配列	129
asymptote (/asymptotic line)	漸近線	84
average	算術平均値	64
axes	両軸	92
axes of symmetry	対称軸	83
axis	軸	87, 93
base	基数	2, 4, 26, 32
	底	32
	底辺	69, 76
	底面	86, 89
binary arithmetic operations	2進算術演算	3
binary notation	2進法表記	2
binary number	2進数	2
binomial	2項式	101
biquadratic	4次	100
biquadratic equation	4次方程式	103
bisect	2等分する	74
braces	中かっこ	21
	{ }	21
brackets	角かっこ	21
	[ ]	21
cancel	消去する	40
cardinal number	カージナル数	18
Cartesian coordinate system	カルテシアン座標系	94
center point	中心点	79

character	文字	1, 3
chord	弦	80
circle	円	79
	円形	87
circular cone	円すい	87
circumference of a circle	円の円周	80
coefficient	係数	99
column	列	129, 130
combination	組合せ	124
common difference	公差	119
common factor	共通因数	40
common fraction	単分数	40
common logarithm	常用対数	32
common ratio	公比	120
common vertex	共通頂点	90
complex fraction	重分数	40
	繁分数	40
	複分数	40
complex number	複素数	16
composite number	合成数	18
cone	すい (錐)	87
congruent	合同な	88
consecutive	連続した	119
consecutive interior angle	連続内角	62
consequent	後項	54
constant	定数	99
constant relation	一定の関係	119
contains as a subset	部分集合として含む	132
	⊃	132
convergent series	収束級数	127
convert	変換する	3, 81
convex regular polyhedron	凸正多面体	88
coordinate axes	座標軸	93
coordinate graph	座標図	92
coordinates	座標	93
corresponding angle	同位角	62

English	Japanese	Page
cosecant	余割	113
cosine	余弦	113
cotangent	余接	113
counting	計数	1
counting number	自然数	13
cube	立方体	88
cube root	3乗根	36
cubic	3次	100
cubic equation	3次方程式	103
curved lateral surface	湾曲した側面	86
cut by	切られる	61
cylinder	円柱	86
decagon	十角形	77
decimal (/decimal fraction)	小数	43, 46
decimal notation	10進法表記	4, 7
decimal number	10進数	3, 11
decimal point	小数点	41
defective number	不足数	18
deficient number	輸数	18
definite integral	定積分	110
degree	度	81
degree of a term	項の次数	99
denominate number	名数	17
denominator	分母	40
depth	奥行き	89
derivative	微分	105
determinant	行列式	129
diagonal	対角線	74
diameter	直径	79
difference	差	20
differential	微分	105
differentiation	微分	105
digit	数字	2, 4, 49
directed number	正負の符号をもった数	18

English	Japanese	Pages
directrix	準線	87
discard	切り捨てる	49
dividend	被除数	16, 21, 40
division	除算	19, 20
divisor	除数	16, 21, 40
	約数	15
dodecahedron	十二面体	88
double integral	2重積分	111
drop	切り捨てる	49
duodecimal notation	12進法表記	3
eccentricity	離心率	84
element	要素	130
ellipse	楕円 (形)	83
elliptical cone	楕円すい	87
endpoint	端点	59
equal	=	22
equal sign	等号	102
equality symbol	等号	102
equation	等式	98, 102, 103
	方程式	98, 102, 103
equiangular (/equilateral) triangle	正三角形	70
even number	偶数	14
exponent	指数	26
	べき指数	26
exponent law	指数法則	27
expression	式	36, 39, 98, 102, 103
exterior angle	外角	62, 70
face angle	面角	88
factor	因数	20
	因数分解する	40
factorial n	$n!$	123
fences	中かっこ	21
	{ }	21

英語索引

English	日本語	ページ
figurate number	多角数	18
figure	数字	4, 49
finite decimal	有限小数	46
focus	焦点	83
formula	公式	84
four basic operations	四則演算	19
fraction (/fractional number)	分数	39
fraction bar	分数バー	20
	———	21
fraction part	小数部	43
function	関数	105, 110
geometric progression (/geometric sequence)	等比級数	120
half line	半直線	59
harmonic progression (/harmonic sequence)	調和級数	122
height	高さ	69, 75, 89
heptagon	七角形	76
hexadecimal notation	16進法表記	3
hexagon	六角形	76
hexagonal prism	六角柱	90
highest degree of any term	各項の最高次数 (変数が1つの場合)	99
hundreds place	100の位	44
hyperbola	双曲線	84
hyperbolic arccosine	逆双曲線余弦	117
hyperbolic arcsine	逆双曲線正弦	117
hyperbolic arctangent	逆双曲線正接	117
hyperbolic cosine	双曲線余弦	116
hyperbolic function	双曲線関数	116
hyperbolic sine	双曲線正弦	116
hyperbolic tangent	双曲線正接	116
icosahedron	二十面体	88

English	日本語	ページ
identity	恒等式	84
ignore	切り捨てる	49
imaginary part of a complex number	複素数の虚数部	17
improper fraction	仮分数	40
indefinite integral	不定積分	110
independent variable	独立変数	105
index	指数	36
inequality	不等式	98, 104
inertial coordinate system	慣性座標系	94
infinite decimal	無限小数	46
infinite number of terms	無限項数	127
integer	整数	13, 14, 41, 43
integer part	整数部	43
integral (/integration)	積分	110
interior angle	内角	62, 69, 73
intersect	交わる	59
intersecting line	交差線	59, 93
intersection of sets	共通集合	132
	∩	132
inverse hyperbolic function	逆双曲線関数	117
inverse of hyperbolic function	双曲線関数の逆元	117
inverse of trigonometric functions	三角関数の逆元	115
inverse trigonometric function	逆三角関数	115
irrational number	無理数	15
is a subset of ～	～の部分集合	132
	⊂	132
is an element of ～	～の要素	132
	∈	132
is equal to (または, equals)	$=$	22
is greater than	$>$	22
is greater than or equal to	$\geq$ (または, $\geqq$)	22
is included in～	～に含まれる	132
is less than	$<$	22
is less than or equal to	$\leq$ (または, $\leqq$)	22

英語索引

is not an element of ～	～の要素ではない	132
	∉	132
is proportional to ～	～に比例	132
	∝	132
is similar to～	～と相似	132
	∽	132
isosceles triangle	2等辺三角形	70
lateral face (/lateral edge)	側面	89
law	法則	84
left side	左辺	102
leg	辺	76
letter	文字	1, 3
like sign	相等しい符号	25
like term	同類項	99
line	線	58
line segment	線分	58, 83
linear	線形	102
linear equation	1次方程式	103
	線形方程式	102
locus	軌跡	85
logarithm	対数	32
lowest term	既約分数	40
major axis	長径	83
matrix	行列	130
mean	算術平均値	64
measuring	計測	1
median	中央値	67
	中線	70
middle number	まん中の数	67
middle two numbers	まん中の2つの数	67
midpoint	中点	70
minor axis	短径	83
minuend	被減数	20
mixed decimal	帯小数	43

English	Japanese	Page
mixed fraction	帯分数	41
mode (/modal class)	最頻値	68
monomial	単項式	100
multiple integral	多重積分	111
multiplication	乗算	19
multiplicative inverse	乗法に関する逆元	25
multiply	かける	81
n decimal places	小数点以下 n けた	44
n factorial	$n!$	123
natural logarithm	自然対数	32
natural number	自然数	13
negation of ~	~の否定	132
negative integer	負の整数	14
negative number	負数	14
No.	番号	2
nonagon	九角形	77
nonlinear equation	非線形方程式	102
nonnegative number	非負数	14
nonnegative real number	非負実数	37
nonparallel	平行でない	76
nonperiodic decimal	非循環小数	47
nonrepeating decimal	非循環小数	47
nonterminating decimal	無限小数	46
normal (/normal line)	法線	84
nth decimal place	小数点以下第 n 位	44
nth-order derivative	n 次微分	106
nth-order partial derivative	n 次偏微分	108
nth power	n 乗	36
nth root	n 乗根	36
nth term	n 次の項	119
number	数	1
	番号	2
number line	数直線	92
number of terms	項数	98

numeral	数字	4
numerator	分数の分子	40
numeric (/numerical)	数字の	10
numeric (/numerical) data	数値データ	11
numeric code	数字コード	11
numeric field	数字フィールド	11
numeric item	数値項目	11
numerical coefficient	数値係数	99
numerical control	数値制御	11
numerical information	数値情報	11
oblique circular cone	斜円すい	87
oblique cylinder	斜円柱	86
oblique prism	斜角柱	89
oblique pyramid	傾斜角すい	90
obtuse angle	鈍角	63
obtuse triangle	鈍角三角形	71
octagon	八角形	77
octahedron	八面体	88
octal notation	8進法表記	3
odd number	奇数	15
omit	切り捨てる	49
on the opposite side of	対向側に	62
on the same side of	同じ側に	62
ones place	1の位	44
opposite angle	対向角度	75
opposite side	対向辺	74
order	次数	129
ordinal number	序数	17
ordinate	縦座標	93
origin	原点	59, 93
oval	楕円	83
parabola	放物線	84
parallel	平行	60

English	Japanese	Page
parallel line	平行線	60
parallelogram	平行四辺形	74
parentheses	かっこ	21
	( )	21
partial derivative	偏微分	108
pattern	図形	9
pentagon	五角形	76
pentagonal prism	五角柱	89
perfect number	完全数	18
periodic decimal	周期小数	47
permutation	順列	123
perpendicular distance	垂直距離	87
perpendicular line	垂線 (または, 垂直線)	59, 69
piece of a line	線の一部分	58
place	位	44
plane	平らな	88
	平面	69, 93
point of intersection	交点	93
polar coordinate system	極座標系	94
polygon	多角形	73
polyhedral angle	多面角	88
polyhedron	多面体	88
polynomial	多項式	16, 98, 100
positive integer	正の整数	14
positive number	正数	14
power	べき	26
	累乗	26
prime (/prime number)	素数	15
principal value	主値	115
prism	角柱	89
product	積	20, 132
	Π	132
progression	級数	119
proper fraction	真分数	40
proportion	比例 (または, 比例算)	55
pure imaginary (number)	純虚数	17

pyramid	角すい	90
Pythagorean theorem (/Pythagoras' theorem)	ピタゴラスの定理	71
quadrangular prism	四角柱	89
quadrant	象限	93
quadratic	2次	100
quadratic curve	2次曲線	84
quadratic equation	2次方程式	103
quadrilateral	四辺形	73, 89
quantity	数量	129
quintic	5次	100
quotient	商	15, 21, 39
radian	ラジアン	80
radical (/radical sign)	根号	36
radicand	被開数	36
radius	半径	79
rate of change	変化率	105
ratio	比	54
	比率	54
	割合	54
rational coefficient	有理数係数	16
rational number	有理数	15
ray	半直線	59
real number	実数	15, 17
real part of a complex number	複素数の実数部	17
reciprocal	逆数	114
rectangle	矩形	74
rectangular array	矩形の配列	130
rectangular coordinate system	直交座標系	92
rectangular solid	長方形立方体	89
recurring decimal	循環小数	46
reduce	通分する	40
regular polyhedron	正多面体	88

remainder	余り	21
repeating decimal	循環小数	46
rhombus	ひし形	75
right angle	直角	59
right circular cone	直円すい	87
right cylinder	直円柱	86
right prism	直角柱	89
right pyramid	直角すい	90
right side	右辺	102
right (-angled) triangle	直角三角形	70, 113
Roman numeral	ローマ数字	2, 9, 11
root	根	16
	乗根	16, 36
round number	端数のない10の倍数	10
rounding down	切り捨てる	53
rounding off	四捨五入する	49
	丸める	49
rounding up	切り上げる	53
row	行	129, 130
scalene triangle	不等辺三角形	70
secant	正割	114
second-order derivative	2次微分	105
second-order determinant	2次行列式	129
sector	扇形	81
semimajor axis	長半径	83
semiminor axis	短半径	83
sequence	数列	119, 122, 127
sequence of numbers	一連の数	119
side	辺	113
sign of equality	等号	21
sign of inequality	不等号	22
signed number	正負の符号をもった数	18
similar term	同類項	99
simple fraction	単分数	40

English	日本語	ページ
sine	正弦	113
single integral	1重積分	110
single-valued function	1価関数	115
slash	/	20
solid figure	立体図形	85
spelled-out number	文字でつづった数字	7
sphere	球	85
square	正方形	74
square face	4角面	88
square number	平方数	18
square root	2乗根	36
statement	記述	104
statistics	統計	64
straight line	直線	58
subtraction	減算	19
subtrahend	減数	20
summation	和 (または, 合計, 総和)	127
surd	無理数	15
surface area of a sphere	球の表面積	85
symbol	記号	1, 21
symmetric	対称性	9
tangent	正接	113
tangent (/tangential line)	接線	84
tens place	10の位	44
term	項	22, 98, 100
terminating decimal	有限小数	46
tetrahedron	四面体	88
therefore	それゆえ	132
	∴	132
thousands place	1000の位	44
transcendental number	超越数	16
transfinite number	超限数	18
transversal	横断線	61
trapezoid	台形	75

triangle	三角形	69
triangular base	三角底辺	89
triangular face	3角面	88
triangular prism	三角柱	89
trigonometric function	三角関数	113
trinomial	3項式	101
triple integral	3重積分	111
truncating	切り捨てる	51
union of sets	和集合	132
	∪	132
unit of a physical quantity	物理量の単位	17
unlike sign	相異なる符号	25
variable	変数	20, 99
vertex	頂点	69, 87
vertical angle	頂角	61
volume of a cone	すいの体積	87
volume of a sphere	球の体積	85
vulgar fraction	単分数	40
weight	重み	66
weighted mean	重みつき平均値	65
whole number	全数 (または, 整数)	13, 15
width	幅	89
x axis	x軸	93
x-coordinate	x座標	93
y axis	y軸	93
y-coordinate	y座標	93

索引

——	fraction bar	21
<	is less than	22
=	equal	22
	is equal to	22
( )	parentheses	21
/	slash	20
[ ]	brackets	21
{ }	braces	21
	fences	21
| |	absolute value bars	21, 132
∼	is similar to ∼	132
∼と相似	is similar to ∼	132
∼に比例	is proportional to ∼	132
∼に含まれる	is included in ∼	132
∼の否定	negation of ∼	132
∼の部分集合	is a subset of ∼	132
∼の要素	is an element of ∼	132
∼の要素ではない	is not an element of ∼	132
>	is greater than	22
∈	is an element of ∼	132
∉	is not an element of ∼	132
∝	is proportional to ∼	132
∠	angle	61
∩	intersection of sets	132
∪	union of sets	132
∴	therefore	132
≤ (または, ≦)	is less than or equal to	22
≥ (または, ≧)	is greater than or equal to	22
⊂	is included in ∼	132

⊃	contains as a subset	132
$n!$	factorial n	123
	n factorial	123
Π	product	132

あ行

アークコサイン	arccosine	115
アークサイン	arcsine	115
アークタンジェント	arctangent	115
相異なる符号	unlike sign	25
相等しい符号	like sign	25
余り	remainder	21
アラビア数字	Arabic numerals	1
ある小数点位以下	after a certain decimal place	49

1次方程式	linear equation	103
(1重)積分	single integral	110
1の位	ones place	44
一連の数	sequence of numbers	119
1価関数	single-valued function	115
一定の関係	constant relation	119
因数	factor	20
因数分解する	factor	40

| 右辺 | right side | 102 |

鋭角	acute angle	63
鋭角三角形	acute triangle	71
$a:b$ (a 対 b)	a to b	54
x 座標	x-coordinate	93
x 軸	x axis	93
n 次の項	nth term	119
n 次微分	nth-order derivative	106
n 次偏微分	nth-order partial derivative	108
n 乗	nth power	36

日本語	English	ページ
n 乗根	nth root	36
円	circle	79
円形	circle	87
円弧	arc	79
円すい	circular cone	87
円すいの体積	volume of a circular cone	87
円柱	cylinder	86
円周	circumference of a circle	80
円の面積	area of a circle	80
扇形	sector	81
横断線	transversal	61
奥行き	depth	89
同じ側に	on the same side of	62
重み	weight	66
重みつき平均値	weighted mean	65

か行

日本語	English	ページ
カージナル数	cardinal number	18
外角	exterior angle	62, 70
角かっこ	brackets	21
各項の最高次数 (変数が1つの場合)	highest degree of any term	99
角すい	pyramid	90
角柱	prism	89
角度	angle	61
かける	multiply	81
加算	addition	19
過剰数	abundant number	18
数	number	1
加数	addend	19
かっこ	parentheses	21
仮分数	improper fraction	40
加法に関する逆元	additive inverse	25
カルテシアン座標系	Cartesian coordinate system	94

関数	function	105, 110
慣性座標系	inertial coordinate system	94
完全数	perfect number	18
記号	symbol	1, 21
記述	statement	104
基数	base	2, 4, 26, 32
奇数	odd number	15
軌跡	locus	85
逆三角関数	antitrigonometric function	115
	inverse trigonometric function	115
逆数	reciprocal	114
逆双曲線関数	inverse hyperbolic function	117
逆双曲線正弦	hyperbolic arcsine	117
逆双曲線正接	hyperbolic arctangent	117
逆双曲線余弦	hyperbolic arccosine	117
既約分数	lowest term	40
球	sphere	85
九角形	nonagon	77
級数	progression	119
球の体積	volume of a sphere	85
球の表面積	surface area of a sphere	85
行	row	129, 130
共通因数	common factor	40
共通集合	intersection of sets	132
共通頂点	common vertex	90
行列	matrix	130
行列式	determinant	129
極座標系	polar coordinate system	94
切られる	cut by	61
切り上げる	rounding up	53
切り捨てる	discard	49
	drop	49
	ignore	49
	omit	49

	rounding down	53
	truncating	51
近似する	approximate	49
偶数	even number	14
矩形	rectangle	74
矩形の配列	rectangular array	130
組合せ	combination	124
位	place	44
傾斜角すい	oblique pyramid	90
係数	coefficient	99
計数	counting	1
計測	measuring	1
弦	chord	80
減算	subtraction	19
減数	subtrahend	20
原点	origin	59, 93
項	term	22, 98, 100
公差	common difference	119
交差線	intersecting line	59, 93
公式	formula	84
項数	number of terms	98
合成数	composite number	18
交点	point of intersection	93
合同な	congruent	88
恒等式	identity	84
項の次数	degree of a term	99
公比	common ratio	120
五角形	pentagon	76
五角柱	pentagonal prism	89
後項	consequent	54
5次	quintic	100
根	root	16

根号	radical (/radical sign)	36

さ行

差	difference	20
最頻値	mode (/modal class)	68
錯外角	alternate exterior angle	62
錯内角	alternate interior angle	62
錯角	alternate angle	62
座標	coordinates	93
座標軸	coordinate axes	93
座標図	coordinate graph	92
左辺	left side	102
三角関数	trigonometric function	113
三角関数の逆元	inverse of trigonometric functions	115
三角形	triangle	69
三角形の面積	area of a triangle	70
三角柱	triangular prism	89
三角底辺	triangular base	89
3角面	triangular face	88
3項式	trinomial	101
3次	cubic	100
3次方程式	cubic equation	103
3重積分	triple integral	111
算術数列	arithmetic sequence	119
算術平均値	arithmetic mean	64
	average	64
	mean	64
3乗根	cube root	36
四角柱	quadrangular prism	89
4角面	square face	88
式	expression	36, 39, 98, 100, 102
軸	axis	87, 93
四捨五入する	rounding off	49
指数	exponent	26
	index	36

次数	order	129
指数法則	exponent law	27
自然数	counting number	13
	natural number	13
自然対数	natural logarithm	32
四則演算	four basic operations	19
実数	real number	15, 17
四辺形	quadrilateral	73, 89
四面体	tetrahedron	88
斜円すい	oblique circular cone	87
斜円柱	oblique cylinder	86
斜角柱	oblique prism	89
周期小数	periodic decimal	47
収束級数	convergent series	127
12進法表記	duodecimal notation	3
十二面体	dodecahedron	88
10の位	tens place	44
重分数	complex fraction	40
16進法表記	hexadecimal notation	3
主値	principal value	115
十角形	decagon	77
10進数	decimal number	3, 11
10進法表記	decimal notation	4, 7
循環小数	recurring decimal	46
	repeating decimal	46
純虚数	pure imaginary (number)	17
準線	directrix	87
順列	permutation	123
商	quotient	15, 21, 39
消去する	cancel	40
象限	quadrant	93
乗根	root	16, 36
乗算	multiplication	20
小数	decimal (/decimal fraction)	43, 45
小数点	decimal point	41

索　引

小数点以下 n けた	n decimal places	44
小数点以下第 n 位	nth decimal place	44
小数部	fraction part	43
焦点	focus	83
乗法に関する逆元	multiplicative inverse	25
常用対数	common logarithm	32
除算	division	19, 20
除数	divisor	16, 21, 40
序数	ordinal number	17
真分数	proper fraction	40
すい (錐)	cone	87
垂線 (または，垂直線)	perpendicular line	59, 69
垂直距離	perpendicular distance	86
数字	digit	2, 4, 49
	figure	4, 49
	numeral	4
数字コード	numeric code	11
数字の	numeric (/numerical)	10
数字フィールド	numeric field	11
数値係数	numerical coefficient	99
数値項目	numeric item	11
数値情報	numerical information	11
数値制御	numerical control	11
数値データ	numeric (/numerical) data	11
数直線	number line	92
数量	quantity	129
数列	sequence	119, 122, 127
図形	pattern	9
正割	secant	114
正弦	sine	113
正三角形	equiangular (/equilateral) triangle	70
整数	integer	13, 14, 15, 41, 43
	whole number	13

正数	positive number	14
整数部	integer part	43
正接	tangent	113
正多面体	regular polyhedron	88
正の整数	positive integer	14
正負の符号をもった数	directed number	18
	signed number	18
正方形	square	74
積	product	132
積分	integral (/integration)	110
接線	tangent (/tangential line)	84
絶対値表示バー	absolute value	21
線	line	58
漸近線	asymptote (/asymptotic line)	84
線形	linear	102
線形方程式	linear equation	102
前項	antecedent	54
全数	whole number	13, 15
線の一部分	piece of a line	58
1000の位	thousands place	44
線分	line segment	58, 83
双曲線	hyperbola	84
双曲線関数	hyperbolic function	116
双曲線関数の逆元	inverse of hyperbolic function	117
双曲線正弦	hyperbolic sine	116
双曲線正接	hyperbolic tangent	116
双曲線余弦	hyperbolic cosine	116
側面	lateral face (/lateral edge)	89
素数	prime (/prime number)	15
それゆえ	therefore	132

た行

対角線	diagonal	74
台形	trapezoid	75

台形の面積	area of a trapezoid	76
対向角度	opposite angle	75
対向側に	on the opposite side of	62
対向辺	opposite side	74
対称軸	axes of symmetry	83
帯小数	mixed decimal	43
対称性	symmetric	9
対数	logarithm	32
代数的数	algebraic number	18
代数的無理数	algebraic irrational number	16
帯分数	mixed fraction	41
平らな	plane	88
楕円 (形)	ellipse	83
	oval	83
楕円すい	elliptical cone	87
多角形	polygon	73
多角数	figurate number	18
高さ	altitude	69, 86, 90
	height	69, 75, 89
多項式	polynomial	16, 98, 100
多重積分	multiple integral	111
縦座標	ordinate	93
多面角	polyhedral angle	88
多面体	polyhedron	88
短径	minor axis	83
単項式	monomial	100
端点	endpoint	59
短半径	semiminor axis	83
単分数	common fraction	40
	simple fraction	40
	vulgar fraction	40
中央値	median	67
中かっこ	braces	21
	fences	21

中心点	center point	79
中線	median	70
中点	midpoint	70
超越数	transcendental number	16
頂角	vertical angle	61
長径	major axis	83
超限数	transfinite number	18
頂点	vertex	69, 87
長半径	semimajor axis	83
長方形立方体	rectangular solid	89
調和級数	harmonic progression (/harmonic sequence)	122
直円すい	right circular cone	87
直円柱	right cylinder	86
直線	straight line	58
直角	right angle	59
直角三角形	right (-angled) triangle	70, 113
直角すい	right pyramid	90
直角柱	right prism	89
直径	diameter	79
直交座標系	rectangular coordinate system	92
通分する	reduce	40
底	base	32
定数	constant	99
定積分	definite integral	110
底辺	base	69, 76
底面	base	86, 89
添加数	adjoined number	18
度	degree	81
同位角	corresponding angle	62
統計	statistics	64

等号	equal sign	102
	equality symbol	102
	sign of equality	21
等差数列	arithmetic progression	119
等式	equation	98, 102, 103
等比級数	geometric progression (/geometric sequence)	120
同類項	like term	99
	similar term	99
独立変数	independent variable	105
凸正多面体	convex regular polyhedron	88
鈍角	obtuse angle	63
鈍角三角形	obtuse triangle	71

な行

内角	interior angle	62, 69, 73
七角形	heptagon	76
2項式	binomial	101
2次	quadratic	100
2次行列式	second-order determinant	129
2次曲線	quadratic curve	84
2次微分	second-order derivative	105
2次方程式	quadratic equation	103
2重積分	double integral	111
二十面体	icosahedron	88
2乗根	square root	36
2進算術演算	binary arithmetic operations	3
2進数	binary number	2
2進法表記	binary notation	2
2等分する	bisect	74
2等辺三角形	isosceles triangle	70

は行

| 配列 | array | 129 |

端数のない数	round number	10
八角形	octagon	77
八面体	octahedron	88
8進法表記	octal notation	3
幅	width	89
半径	radius	79
番号	No.	2
	number	2
半直線	half line	59
	ray	59
繁分数	complex fraction	40
比	ratio	54
被開数	radicand	36
被減数	minuend	20
ひし形	rhombus	75
非循環小数	nonperiodic decimal	47
	nonrepeating decimal	47
被除数	dividend	16, 21, 40
非線形方程式	nonlinear equation	102
ピタゴラスの定理	Pythagorean theorem	71
	(/Pythagoras' theorem)	
非負実数	nonnegative real number	37
非負数	nonnegative number	14
微分	derivative	105
	differential	105
	differentiation	105
100の位	hundreds place	44
比率	ratio	54
比例（または，比例算）	proportion	55
複素数	complex number	16
複素数の虚数部	imaginary part of a complex number	17
複素数の実数部	real part of a complex number	17
複分数	complex fraction	40

負数	negative number	14
不足数	defective number	18
物理量の単位	unit of a physical quantity	17
不定積分	indefinite integral	110
不等号	sign of inequality	22
不等式	inequality	98, 104
不等辺三角形	scalene triangle	70
負の整数	negative integer	14
部分集合として含む	contains as a subset	132
分数	fraction (/fractional number)	39
分数の分子	numerator	40
分数バー	fraction bar	21
分母	denominator	40
平行	parallel	60
平行四辺形	parallelogram	74
平行線	parallel line	60
平行でない	nonparallel	76
平方数	square number	18
平面	plane	69, 93
べき	power	26
べき指数	exponent	26
辺	leg	76
	side	113
変化率	rate of change	105
変換する	convert	3, 81
変数	variable	20, 99
偏微分	partial derivative	108
豊数	abundant number	18
法線	normal (/normal line)	84
法則	law	84
方程式	equation	98, 102, 103
放物線	parabola	84
ほぼ等しい	approximately equal	16

ま行

交わる	intersect	59
丸める	rounding off	49
まん中の数	middle number	67
まん中の2つの数	middle two numbers	67
無限項数	infinite number of terms	127
無限小数	infinite decimal	46
	nonterminating decimal	46
無名数	absolute number	17
無理数	irrational number	15
	surd	15
名数	denominate number	17
面角	face angle	88
文字	character	1, 3
	letter	1, 3
文字でつづった数字	spelled-out number	7
約数	divisor	15
有限小数	finite decimal	46
	terminating decimal	46
有理数	rational number	15
有理数係数	rational coefficient	16
要素	element	130
余割	cosecant	113
余弦	cosine	113
横座標	abscissa	93
4次方程式	biquadratic equation	103
余接	cotangent	113
4次	biquadratic	100
ラジアン	radian	80

離心率	eccentricity	84
立体図形	solid figure	85
立方体	cube	88
両軸	axes	92
輪数	deficient number	18
隣接角	adjacent angle	61
累乗	power	26
列	column	129, 130
連続した	consecutive	119
連続内角	consecutive interior angle	62
ローマ数字	Roman numeral	2, 9, 11
六角形	hexagon	76
六角柱	hexagonal prism	90
和 (または, 合計, 総和)	summation	127
y座標	y-coordinate	93
y軸	y axis	93
和集合	union of sets	132
割合	ratio	54
湾曲した側面	curved lateral surface	86

1

数と数を表す数字と文字
number and digit/figure/numeral

数   number

「数」(number) は,計数 (counting) または計測 (measuring) のために使用され,一般に「数字」,「記号」または「文字」(characters または letters) などを使用して表されます.

● digit (any of the numbers from 0 to 9)

0から9までの数,すなわち数字.われわれが日常使用している10進数を含み,そのほかデジタル技術分野で広く使用されている2, 8, 12および16進数などにも使用されます.

● figure (any of the number signs from 0 to 9)

0から9までの数,すなわち数字で,主として10進数を表す数字として使用されています.

● numeral (a sign that represents a number)

数を表す記号のことで,この中には主として,10進数の「数字」の意味で使用される0〜9 (アラビア数字: Arabic numerals) と「記号」が含まれます.数を表す「記号」(symbol: a sign which expresses a number) としては,主として

ローマ数字が使用されます．

　Roman numeral (ローマ数字) [any of the signs (such as I, II, V, X, L, C, etc.) used for numbers in ancient Rome and sometimes now]

注1)　number は，「数」以外に，「ある人または物を識別するために使用される，一群の数 (a group of numbers that is used to identify somebody/something)」を表す用語としても使用され，このとき日本語訳は「番号」または「ナンバー」となります．

　この場合，ときには number(s) のかわりに No.(複数形は Nos.)，no (複数形は nos.) また，米国英語では記号 # が使用されることがあります．

　　部屋番号 347：Room No. (または，number) 347
　　番地：Address No. (または，number)
　　製造一連番号 (または，製造番号)：
　　　　Manufacturer's serial number

注2)　2進法表記 (binary notation)
　2進数 (binary numbers) とは2進法表記 (binary notation) で表される数のことで，基数 (base) 2 を使用する1つの数体系 (a number system using the base two) のことです．
　2進数を表すために，2個の数字 (digits)，0 および 1 が使用されます．2進法表記では 0 と 1 以外の数字は使用しません．

デジタルコンピュータは2進法にもとづいてすべての計算を行うので，2進法はコンピュータによる計算には非常に便利な方法です．

コンピュータは10進数を2進数に変換して (to convert)，決められたプログラムにしたがって2進算術演算［binary arithmetic operations: addition/subtraction/multiplication/division］を行い，得られた結果を再度10進数に変換して，表示および印字を行います．

10進数 (decimal number)	2進数 (binary number)
0	0
1	1
2	10
3	11
4	100
5	101
6	110
7	111
8	1 000
9	1 001
10	1 010
16	10 000

そのほかの表記 (other notations)

8進法表記 (octal notation) では8個の数字 (eight digits)，すなわち，0, 1, 2, 3, 4, 5, 6, 7を使用し，12進法表記 (duodecimal notation) では，10個の数字 (ten digits)，0, 1, 2, 3, 4, 5, 6, 7, 8, 9と2と3を逆に書いた2つの文字，Ɫ, Ɛ (characters，または，letters) (2 and 3 written upside down: ten is Ɫ and eleven is Ɛ) が使用され，16進法表記 (hexadecimal notation) では10個の数字 (ten digits), 0, 1, 2, 3, 4, 5, 6, 7, 8, 9と6個の文字 (6 characters，または，letters), A, B, C,

D, E, Fを使用します.

本書では, われわれが日常生活で使用する, 基数 (base) が 10, すなわち 0 から 9 の 10 個の数字の組合せで「数」を表す, 10 進法表記 (decimal notation) について説明します.

数字   `digit, figure, numeral`

日本語の「数字」に相当する英語としては, digit, figure および numeral があります.

● digit

0 から 9 までのいずれかの数字 (any of the numbers from 0 to 9).

一般には, デジタル技術分野の用語として使用されます. 例えば 2 進法では 0, 1 の 2 個の数字 (two digits), 8 進法では 0 ～ 7 の 8 個の数字 (eight digits) および 10 進法, 12 進法, 16 進法では 0 ～ 9 の 10 個の数字 (ten digits) のことです.

構文例および読み方

◯ 2 進法で数を書く場合には, 10 進法で数を書く場合と比較して多くの数字が必要です.

➡ Writing numbers in binary requires more digits than writing numbers in decimal.

⊃ 数 2005 には 4 つの数字が含まれます.

➡ The number 2005 contains four digits.

⊃ 別の循環小数の例は
$$\frac{5}{7}$$
で，これは小数で
 0.714 285 714…
と表されます．これは，714 285 の数字ブロックをもっています．

➡ Another example of a repeating decimal is
$$\frac{5}{7},$$
which is
 0.714285714…
with the block of digits 714285.

注) digit の本来の意味は，0 から 9 までのいずれかの数字 (any of the numbers from 0 to 9) ですが，日本語で「桁」と訳したほうが，意味が明確になる場合が多々あります．

例えば，

Whole numbers (consisting) of five digits or more (but not decimal fractions) use a comma to separate three-digit group, counting from the right (e.g. 1,500,000).

・5 個，またはそれ以上の個数の数字からなる整数 (ただし小数は除く) は，カンマを使用して，右側から数えて 3 個の数字のグループごとに分けます (例えば 1,500,000).

Figures of four digits (Four-digit figures) may be written without a comma (e.g. 2000 instead of 2,000).

・4個の数字からなる数字は，カンマを省略して書いてもかまいません (例えば 2,000 のかわりに 2000).

とそれぞれ訳すことができますが，これでも日本語として理解可能です．

これらの場合，digits を「数字」ではなく「桁」と訳すと，日本語としての意味が明確になります．

すなわち，それぞれ
・5桁またはそれ以上の桁数の整数 (ただし小数は除く) は，カンマを使用して，右側から数えて3桁ごとにグループ分けします (例えば 1,500,000).
・4桁の数字では，カンマを省略して書いてもかまいません (例えば 2,000 のかわりに 2000).

となります．下記は，次のように訳せます．

Five or more digits to the right of the decimal point is sometimes separated into three-digit groups by a thin space, counting from the decimal point (e.g. 209.775⎵42).

・小数点から右側の5桁またはそれ以上の桁数は，ときには小数点から数えてわずかなスペースを空けて，3桁のグループごとに分けることもあります (例えば 209.775⎵42).

The number of digits is stored in a separate location.

・桁数は別の場所に記憶されます.

● figure

0 ～ 9 の数字のいずれか (any of the number signs from 0 to 9) です. figure は，一般に，われわれが日常生活で取り扱う 10 進法表記 (decimal notation) で使用されます.

構文例および読み方

⊃ 主な問題は，数を文字でつづるか，数字で表すかです.

➡ The major issue is whether to spell out numbers or to express them in figures.

⊃ 一般に，数字で書いたほうが，文字でつづった数より読みやすくなります.

➡ Figures are generally easier to read than spelled-out numbers.

⊃ 5 桁の数字はカンマをつけて書きます (例えば 32,000).

➡ Figures of five digits (Five-digit figures) are written with a comma (e.g. 32,000).

注 1) figure も digit 同様，0 ～ 9 の数を表す数字のいずれかの意味ですが，「数字」と訳すよりは「桁」と訳したほうがわかりやすい場合が多々あります.

His income is in six figures per month.

(He has a six-figure income per month.)
・彼の収入は1か月あたり6個の数字です．

が本来の意味ですが，これを「彼の収入は1か月あたり6桁の数字です」とすればさらに意味がはっきりします．すなわち，円での収入であれば，1か月あたり少なくとも100,000円であることを表しています．

上文はまた，次のように書き換えても意味は同じです．
His income is in six digits per month.
(He has a six-digit income per month.)

注2) figure は，ある物を説明するために，書面中に使用される番号づけされた図面や図 (a numbered drawing or diagram used in a book to explain something)，と訳される場合もあります．

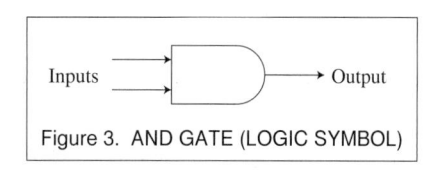

Figure 3. AND GATE (LOGIC SYMBOL)

・図3はANDゲートを表しています．
Figure 3 shows the AND gate.

Figure 3 の Figure を Fig.と省略する場合もあります．

・図10, 11は，本装置の操作を説明するためのものです．
　Figures (Figs.) 10 and 11 are for explaining the operation of this equipment.

図が複数の場合は Figures となります.

注3) figure は，図形 (a pattern) と訳される場合もあります.

・一般に，全体をかえずに図形の一部分を交換できるとき，対称性があるといわれます.

In general, a figure is said to be symmetric if parts of it may be interchanged without changing the whole.

● numeral

数を表す記号 (a sign that represents a number) のことです.

numeral は，上記の説明のとおり「数」を表す記号ですから，むろん 0～9 の数字だけでなく，それ以外の記号でもかまわないわけです．したがって，numeral は figure や digit のかわりに使用できます.

10 進数表記では，数字を表す numerals, digits, figures が混在して使用される場合が多々あります.

・ローマ数字 (Roman numerals)

詳細はローマ数字表記を参照.

構文例および読み方

⬤ 4桁の数字が表の一部を形成する場合，これら4桁の数字を5桁，またはそれ以上の桁の数字とそろえることができるようにするために，カンマが必要です．

➡ If numerals of four digits form part of a tabulation, commas are necessary so that the numerals can align with numerals of five or more digits.

○ ちょうどの数は文字表示されずに,数字で表されます (例えば千のかわりに 1,000).

➡ Round numbers* are not spelled out but are expressed as numerals (e.g. 1,000 instead of one thousand).
   * 10, 100, 1000 など,10 の倍数で表される端数のない数のことです.

○ 技術文章では,単位つきのすべての数は,数字で書かれます (例えば,30 m).

➡ In technical writing, all numbers used with units of measurement are written as numerals (e.g. 30 m).

注1) 上記の文中すべての numerals を figures にかえることができます.

注2) 「数字の」という形容詞形で使用する際は,numeric または numerical を使用します.

構文例および読み方

○ たいていのキーボードは,右側に数字キーパッド (数字打ちこみ用) がついています.

➡ Most keyboards have a numeric keypad (for typing digits) at the right.

そのほか，下記の表現方法が使用されます．

数字コード: numeric code
数値データ: numeric (numerical) data
数値項目: numeric item
数字フィールド: numeric field
数値制御: numerical control
数値情報: numerical information

ローマ数字表記
(notation of numbers expressed as Roman numerals)

ローマ数字と10進数の関係

前述したとおり，一般に「数」は，0〜9の数字およびA〜Eなどの文字を使用して表記されますが，ときに10進数ではI, II, III, V, Xなどのローマ数字 (Roman numerals) を使用して表記されることがあります．ローマ数字と10進数の関係は次のとおりです．

ローマ数字 (Roman numeral)	10進数 (decimal number)	ローマ数字 (Roman numeral)	10進数 (decimal number)
I (i)	1	MM (mm)	2,000
II (ii)	2	$\overline{\text{V}}$ (v)	5,000
III (iii)	3	$\overline{\text{X}}$ (x)	10,000
IV (iv)	4	$\overline{\text{C}}$ (c)	100,000
V (v)	5	$\overline{\text{M}}$ (m)	1,000,000
X (x)	10		
L (l)	50		
C (c)	100		
D (d)	500		
M (m)	1,000		

これらローマ数字の配列は，値の高いものに順次，低いものを後続させます．
(例)　MDCLXVI＝1,666

2 数の集団
groups of numbers

代表的な数の集団

10進数は算術計算を行う上で,いくつかの数の集団に分類されます.以下がその代表例です.

- 自然数:natural,または,counting number
  1, 2, 3, …のような正の整数.
  整数 (integer) を参照.

- 全数* (または整数):whole number
  ゼロを含むすべての自然数.

  (例) 0, 1, 2, 3, …

  * 全数は整数に含まれるので,一般には,全数とは訳さずに整数と訳されています.

- 整数:integer
  0を含む正および負の全数 (whole numbers).

  (例) 0, ±1, ±2, ±3, …

注) 整数は integer と呼び，integer number とはいいません．

- **正の整数：positive integer**

  自然数 (natural numbers) を正の整数とも呼びます．

  (例)　1, 2, 3, …

- **負の整数：negative integer**

  ゼロを含まない負の全数．

  (例)　$-1, -2, -3, \cdots$

- **正数：positive number**

  ゼロより大きい (ゼロを含まない) 実数 (real numbers).

  (例)　$1, 3, \sqrt{10}, \cdots$

  実数：real number を参照．

- **負数：negative number**

  ゼロ以下の (ゼロを含まない) 実数 (real numbers).

  (例)　$-1, -3, -\sqrt{10}, \cdots$

  実数：real number を参照．

- **非負数：nonnegative number**

  ゼロに等しいか，またはゼロより大きい実数．

- **偶数：even number**

  2 で割ることのできる整数 (integers).

  (例)　$\pm 2, \pm 4, \pm 6, \cdots$

- **奇数**：odd number
  2で割ることのできない整数 (integers)．

  (例)　±1, ±3, ±5, …

- **素数**：prime，または，prime number
  1より大きい (1を含まない) 全数 (whole numbers) で，1およびそれ自身以外に約数 (divisor) をもたない数．

  (例)　2, 3, 5, 7, …, 101, …, 1093, …

- **実数**：real number
  有理数 (rational numbers) と無理数 (irrational numbers) の総称．

  (例)　$-\dfrac{2}{5}, \dfrac{1}{2}, \sqrt{2}, 6, \cdots$

- **有理数**：rational number
  整数，または2つの整数の商 (quotient) として書くことのできる実数．

  (例)　$-1.33, 1, 7, 540, \dfrac{2}{3}, \dfrac{1}{9}, \cdots$

- **無理数**：irrational number (または，surd)
  整数，または2つの整数の商 (quotient) として書くことのできない実数．

  $$q = a \div b$$

  ここで，

  　q：商 (quotient)

2　数の集団

groups of numbers

a : 被除数 (dividend)
b : 除数 (divisor)

無理数には2種類あります.

(i) 代数的無理数 : algebraic irrational number
有理数係数 (rational coefficient) をもつ多項式 (polynomial equation) の根* (root) となる無理数.

* 与えられた多項式の変数に代入された数が,その多項式を満足する場合,この数を根 (root) と呼びます.
 一方,ある数が累乗された場合,与えられた数となるその数を乗根 (root) と呼びます.すなわち英語では同じですが,日本語の呼び名がそれぞれ異なります.
 (例) $\sqrt{5}$ ($\fallingdotseq$**2.236) ($\sqrt{5}$ は $x^2-5=0$ の根で, 2.236 が 5 の 2 乗根です.)
** $\fallingdotseq$ は,「ほぼ等しい (is approximately equal to~)」を表す記号です. $\fallingdotseq$ と同様の意味で, $\approx$ も使われます.

(ii) 超越数 : transcendental number
有理数係数をもつ,多項式の根とはならない無理数.
 (例) π, または, e

● 複素数 : complex number

$z = a + ib$ で表される数.
 ここで,
  i : $\sqrt{-1}$

a, b : 実数 (real numbers)
a : 複素数の実数部
(the real part of a complex number)
b : 複素数の虚数部
(the imaginary part of a complex number)

$b=0$ なら，$z=a$ となり，すなわち虚数部が0となり，実数となります．

$a=0$ で $b \neq 0$ のとき，$z=ib$ となり，純虚数[pure imaginary (number)] と呼びます．

複素数は，実数部と虚数部を別々に加算 (減算) できます．

● 序数 : ordinal number
順序を示す数．
(例) first, second, third, 15th など．

● 名数 : denominate number
物理量の単位 (a unit of a physical quantity) を決定する数．
(例) 5 m や 6 V の，5 や 6．

そのほかの数の集団

上記の代表的な数の集団のほかにも，次の数の集団が存在します．

無名数 : absolute number

豊数，または，過剰数：abundant number
添加数：adjoined number
代数的数：algebraic number
カージナル数：cardinal number
合成数：composite number
輪数 (または，不足数)：deficient number
　　　　　　　　　　(または，defective number)
正負の符号をもった数：directed number
　　　　　　　　　　(または，signed number)
多角数：figurate number
完全数：perfect number
平方数：square number
超限数：transfinite number

3 四則演算
four basic operations

四則演算

四則演算 (four basic operations) とは，加算 (addition)，減算 (subtraction)，乗算 (multiplication) および除算 (division) のことです．

- 加算 (addition)，記号：＋

(例)　　　　　　　　　　読み方

$a + b = c$
- The sum of a and b is c.
- a plus b equals c.
- a added to b equals c.

a および b を加数 (addends)，c は和 (sum) と呼びます．

- 減算 (subtraction)，記号：−

(例)　　　　　　　　　　読み方

$a - b = c$
- The difference of a and b is c.
- a minus b equals c.
- a less b equals c.
- b subtracted from a equals c.

a を被減数 (minuend), b を減数 (subtrahend), c を差 (difference) と呼びます.

● **乗算** (multiplication), 記号：×, または, ・

注1) 乗算を表す× が文字の x と混同するおそれのある場合, × のかわりに ・ を使用します.

注2) 変数 (variables) 間, または数と変数間に何も符号がない場合には, 乗算を意味します.
(例)
$ab \to a \cdot b$ (または, $a \times b$)
$3x \to 3 \cdot x$ (または, $3 \times x$)

ただし, 2つの数字を並べる場合, 例えば 34 は数 34 の意味です. 3・4 (または, 3 × 4) の意味ではありません.

注3) a と b の乗算の場合, $a(b)$, $(a)b$ または $(a)(b)$ と書いてもかまいません.

(例)　　　　　　　　　　読み方
$a \cdot b = c$ 　　　　　　・ The product of a and b is c.
(または, $a \times b$)　　　・ The product of a multiplied by b is c.
　　　　　　　　　　　　・ a times b is c.

a と b を因数 (factors), c を積 (product) と呼びます.

● **除算** (division), 記号：÷, または, ―*, /**, $\sqrt{}$)
* ― を fraction bar と呼びます.
** / を slash と呼びます.

3　四則演算
four basic operations

(例)	読み方
$a \div b = c$	· The quotient of a and
(または, $\dfrac{a}{b} = c$)	b is c.
($a/b = c$, $b\overline{)a} = c$)	· a divided by b equals c.
	· a over b equals c.

a を被除数 (dividend), b を除数 (divisor), c を商 (quotient) と呼びます. 除算の結果, 割り切れない場合は余り (remainder) が生じます.

注4) 四則演算には次の記号が使用されます.

記号 (symbol)	名　称
( )	parentheses (かっこ)
[ ]	brackets (角かっこ)
{ }	braces (または, fences) (中かっこ)
| |	absolute value bars (絶対値表示バー)
────	fraction bar (分数バー)

注5) 分数バーを除く記号には, それぞれ複数形が使用されます. 片側1個でよい場合は, それぞれ単数形となります.

注6) かっこ使用の順番としては最初に ( ), 次に [ ], それから { } となります.

$$\{[(\quad)]\}$$

注7) 等号, または不等号

いままでの説明では, $a+b=c$ のように等号 (the sign of

equality) を使用して説明しましたが，この等号を不等号 (the sign of inequality) に置き換えることができます．

記号	読み方	例
=	equals (is equal to)	$a = b$
<	is less than	$a < b$
>	is greater than	$a > b$
≤ (≦)	is less than or equal to	$a \leq b$
≥ (≧)	is greater than or equal to	$a \geq b$

構文例および読み方

◯ $2(9-7)$

➡ 2 times (multiplied by) 9 minus 7, both terms* in the parentheses.
(2 on the outside of the parentheses times 9 minus 7 on the inside.)

* terms (項) については式の項を参照．

◯ $(7-3)+(6-4)=6$

➡ 7 minus 3, both terms in(within, または, enclosed in) the parentheses plus 1** on the outside of the parentheses times 6 minus 4 on the inside equals 6.

** $+(6-4)$は$+1(6-4)$のことです．

◯ $7 \cdot 3 + 2(4-1) = 27$

➡ 7 times 3 plus 2 on the outside of the parentheses times 4 minus 1 on the inside equals 27.

◐ $(5)(-4)(-9)$

➡ 5 times minus 4 times minus 9.

◐ $\frac{2^3}{4}+4\,(7-2)=22$

➡ 2 cubed divided by 4, the divisor plus 4 on the outside of the parentheses times 7 minus 2 on the inside equals 22.

◐ $2[4\,(2+1)+3]=2[12+3]=30$

➡ 2 on the outside of the brackets times 4 on the outside of the parentheses times 2 plus 1 on the inside of the parentheses plus 3 on the inside of the brackets equals 2 on the outside of the brackets times 12 plus 3 on the inside and also equals 30.

◐ 次の式を簡略化しなさい．
$$\frac{2\,(6-3)+8}{2}=7$$

➡ Simplify the following expression.
2 multiplied by 6 minus 3, both terms in the parentheses plus 8 divided by 2 equals 7.

◐ $(x-3)\,(x-7)$

3　四則演算
four basic operations

➡ x minus 3, both terms in the parentheses multiplied by x minus 7, both terms in the parentheses.

⭗ 11 を 3 で割れば余りは 2 となります．

➡ If you divide 11 by 3, the remainder is 2.
(When dividing 11 by 3, the remainder is 2.)

⭗ 次の式のかっこを開きなさい．

$$4(x-2)+3x=7x-8$$

➡ Remove the parentheses from the following expression.
4 on the outside of the parentheses times x minus 2 on the inside plus $3x$ equals $7x$ minus 8.

⭗ $0 > -2$

➡ 0 is greater than minus 2.

⭗ $-6 < 1$

➡ Minus 6 is less than 1.

⭗ $5 \neq 2$

➡ 5 is not equal to 2.

⭗ 次の記述が真となるように，<（または，>）を挿入しなさ

> い.
>
>   3 6 → 3 < 6
>
> ➡ Insert < or > to make the following statement true.
>
>   3 6
>
> The answer becomes as follows.
>   3 < 6

注8) 四則演算では，以下の用語もしばしば使用されます．

・相等しい符号：like sign
例えば +5 と +2，−5 と −2．それぞれ相等しい符号をもっています．

・相異なる符号：unlike sign
例えば +5 と −2，−5 と +2．それぞれ相異なる符号をもっています．

・加法に関する逆元：additive inverse
減算は加算としても定義されます．
例えば，5−3 は 5 +(−3) とも書けます．ここで，−3 は 3 の加法に関する逆元を表しています．

・乗法に関する逆元：multiplicative inverse
積が 1 となる 2 つの数の一方を，他方に対して，乗法に関する逆元といいます．

例えば，

10, と, $\frac{1}{10}$ ($10 \times \frac{1}{10} = 1$)

の一方がそうです．

指数，べき指数   exponent

ある数，または変数の右側の上付き数で，乗算の回数を示しています．

(A number placed in a superscript position to the right of another number or variable to indicate repeated multiplication.)

例えば，
$$5^3 = 5 \times 5 \times 5 = 125$$
で，5 を基数 (base)，3 を指数 (exponent) と呼びます．

(In 5^3, the number 5 is the base and the number 3 is the exponent.)

累乗，べき   power

ある数がそれ自身により乗算される回数です．

(the number of times that a number is to be multiplied by itself.)

power のかわりに exponent ということもできます．

例えば 2 の 3 乗は 2^3 と書かれ，$2 \times 2 \times 2$ を意味します．

(The base 2 to the power of 3 is written 2^3, and means $2 \times 2 \times 2$.)

注) より厳密に exponent と power を区別すると，power は同じ数を何回かかけ算して得られた結果を意味します．

累乗 (または指数) の基本の読み方は次のとおりです．

累乗 (または指数)	読み方
3^0	: 3 to the zero power
$3^2 \ (=3\times 3)$	: 3 squared，または，3 to the second power*
$4^3 \ (=4\times 4\times 4)$	: 4 cubed，または，4 to the third power
$2^4 \ (=2\times 2\times 2\times 2)$	: 2 to the fourth power
$2^n \ (=2\times 2\times \cdots \times 2)$	: 2 to the nth power
$2^{-n}\left(=\dfrac{1}{2^n}=\dfrac{1}{2\times 2\times 2\times \cdots \times 2}\right)$	: 2 to the minus nth power

* 一般に 3 to the second exponent とはいいません．

以下に，代表的指数法則 (exponent laws) について記述します．

かけ算の場合，基数が同じならそれぞれの指数を加算します．
(When multiplying, if the bases are the same, add the exponents.)
$$a^m \cdot a^n = a^{m+n}$$
(例)
$2^3 \, 2^2 = 2^{3+2} = 2^5 = 2\times 2\times 2\times 2\times 2 = 32$

累乗の累乗は，それぞれの指数を乗算します．
(When raising a power to a power, multiply the powers.)

$$(a^m)^n = a^{mn}$$

(例)

$(2^3)^2 = 2^6 = 2 \times 2 \times 2 \times 2 \times 2 \times 2 = 64$

積の累乗は各因数を累乗します．

(When raising a product to a power, raise each factor to the power.)

$$(a \cdot b)^m = a^m b^m$$

(例)

$(2 \cdot 3)^2 = 2^2 \cdot 3^2 = 2 \times 2 \times 3 \times 3 = 36$

分数の累乗は，分子と分母を累乗します．

(When raising a fraction to a power, raise the numerator and denominator to the power.)

ただし，$b \neq 0$ です．

$$\left(\frac{a}{b}\right)^m = \frac{a^m}{b^m}, \text{ where } b \neq 0$$

$$\left(\frac{2}{3}\right)^2 = \frac{2^2}{3^2} = \frac{4}{9} = 0.\dot{4}\text{*}$$

＊ "・" については，小数の項を参照 (37 ページ)．

除算の場合，基数が同じ場合は，分母の指数を分子の指数から引きます．

(When dividing, if the bases are the same, subtract the exponent in the denominator from the exponent in the numerator.)

ただし，$a \neq 0$ です．

$$\frac{a^m}{a^n} = a^{m-n}, \text{ where } a \neq 0$$

(例)

$$\frac{2^3}{2^2} = 2^{3-2} = 2^1 = 2$$

ゼロ以外の数のゼロ乗は1となります．

(A nonzero number raised to the zero power equals 1.)

$a^0 = 1$, where $a \neq 0$

(例)

$2^0 = 1$

ゼロ以外の数の $-n$ 乗は，1をその数の n 乗で割った数と等しくなります．

(A nonzero number raised to the minus nth power equals 1 divided by the number raised to the nth power.)

$a^{-n} = \dfrac{1}{a^n}$, where $a \neq 0$

(例)

$2^{-2} = \dfrac{1}{2^2} = \dfrac{1}{4} = 0.25$

基本の読み方は下記のとおりです．

$(2^3)^2$: Raise 2 to the third power to the second power.

$(2 \cdot 3)^3$: Raise the product of 2 and 3 to the third power.

$\left(\dfrac{2}{3}\right)^2$: Raise the fraction of $\dfrac{2}{3}$ to the second power.

構文例および読み方

● $4^2 \cdot 4^3$

➡ 4 squared times 4 cubed.

◯ $2^{-3} = \dfrac{1}{2^3} = \dfrac{1}{8}$

➡ 2 to the minus third power equals one divided by 2 to the third power and also equals one-eighth.

◯ $x^4 \cdot x^2 = x^6$

➡ x to the fourth power times x squared equals x to the sixth power.

◯ $(x^2)^3 = x^6$

➡ x raised to the second power to the third power equals x to the sixth power.

◯ $(3x^2 + 4x + 7) + (5x^2 + 2x + 12)$

➡ 3 times x squared plus $4x$ plus 7, these three terms in the parentheses plus 1 on the outside of the parentheses times 5 times x squared plus $2x$ plus 12 on the inside.

◯ $\dfrac{12a^6 - 8a^4 + 7a}{2a^2} = 6a^4 - 4a^2 + \dfrac{7}{2a}$

➡ 12 times a to the sixth power minus 8 times a to the fourth power plus $7a$ divided by 2 times a squared equals 6 times a to the fourth power minus 4 times a squared plus 7 divided by $2a$.

◯ 4 は 2 の 2 乗です（$=2^2$）．

➡ 4 is the second power of 2 ($=2^2$).

3 四則演算
four basic operations

- 2^5 の値を求めなさい．ここで 2 は基数と呼ばれ，5 は指数と呼ばれます．

→ Find the value of 2^5, where 2 is called the base, and 5 is called the exponent.

- 2 と 3 の積は 6 です．

→ The product of 2 and 3 is 6.

次のように書き換えることもできます．
The product of 2 multiplied by 3 is 6.

- (a) $(4x)^0 = 4^0 x^0 = 1 \cdot 1 = 1$
  (b) $4x^0 = 4 \cdot 1 = 4$

  (a)では積 $4x$ が 0 乗され，一方 (b) では x のみが 0 乗されています．

→ In (a), the product $4x$ is raised to the zero power, while in (b) only x is raised to the zero power.

- 3 を 2 乗します．

→ Raise 3 to the second power.

- 2 の 3 乗 ($=2^3$) は 8 です．

→ 2 raised to the third power is 8.

対数　　　　　　　　　　　　　　　`logarithm`

$y = a^x$ の式で，y を整数とした場合，これを対数式で表すと次のようになります．

$x = \log_a y$

上式で a を基数 (または，底) (base) と呼びます．

基数が 10 の対数を常用対数 (common logarithm：$\log_{10} y$)，基数が e ($=2.718\cdots$) の対数を自然対数 (natural logarithm：$\log_e y$) と呼びます．また基数が 10 の場合，基数 10 を省略できます．

$\log_{10} y = \log y$

一般に常用対数と自然対数の間には次の関係があります．

$\log_{10} x = M \times \log_e x$

$\log_e x = M^{-1} \times \log_{10} x$

(ここで，$M = \log_{10} e = 0.433\,29\cdots$)

$M^{-1} = 2.302\,58\cdots$

$\log_e x$ は $\ln x$ とも書かれます．

自然対数を使用すると，微分および積分がより簡単に扱えます．

以下に，対数に適用される代表的な法則について記述します．下記の数 n および m は正数です．

基数が a のとき，a の対数は 1 です．
(The logarithm of a to the base a equals 1.)

$\log_a a = 1$

(例)

$\log_{10} 10 = 1$

基数が a のとき，1 の対数は 0 です．

(The logarithm of 1 to the base a equals 0.)

$\log_a 1 = 0$

(例)

$\log_{10} 1 = 0$

基数が a のとき，$n \cdot m$ の対数は，基数 a の n の対数と，基数 a の m の対数の和に等しくなります．

(The logarithm of $n \cdot m$ to the base a equals the logarithm of n to the base a plus the logarithm of m to the base a.)

$\log_a (n \cdot m) = \log_a n + \log_a m$

(例)

$\log_{10} 50 = \log_{10}(5 \times 10) = \log_{10} 5 + \log_{10} 10$
$\qquad = 1 + \log_{10} 5 \ (= 0.699\,0 :\ 常用対数表より)$
$\qquad = 1 + 0.699\,0 = 1.699\,0$

基数 a の n/m の対数は，基数 a の n の対数から，基数 a の m の対数を引いた値です．

(The logarithm of n/m to the base a equals the logarithm of n to the base a minus the logarithm of m to the base a.)

$\log_a \left(\dfrac{n}{m} \right) = \log_a n - \log_a m$

(例)

$\log_{10} \dfrac{1}{10} = \log_{10} 1 - \log_{10} 10$
$\qquad = 0 - 1 = \bar{1}\ $ (where $\bar{1}$ is read as 'bar one.')

基数 a の n の m 乗である対数は，m に基数 a の n の対数を乗算した値です．

(The logarithm of n to the mth power to the base a equals m times the logarithm of n to the base a.)

$$\log_a n^m = m \log_a n$$

(例)

$$\log_{10} 10^3 = 3 \log_{10} 10 = 3 \times 1 = 3$$

注) 記号，文字または数字の上付き，下付き文字の表現は，次のとおりです．

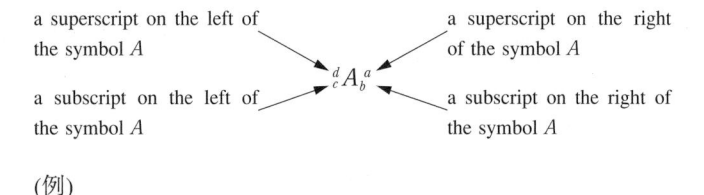

a superscript on the left of the symbol A → $^d_c A^a_b$ ← a superscript on the right of the symbol A

a subscript on the left of the symbol A → $^d_c A^a_b$ ← a subscript on the right of the symbol A

(例)

a_n : a with the subscript n on the right

構文例および読み方

● $\log(n \cdot m) = \log(n) + \log(m)$

➡ The logarithm of n times m equals the logarithm of n plus the logarithm of m.

● $\log n^m = m \log(n)$

➡ The logarithm of n to the mth power equals m times the logarithm of n.

○ $\log_{10} x = M \log_e x$

➡ The logarithm of x equals M times the logarithm of x to the base e.

○ 100 の常用対数は 2 です．というのは，$100 = 10^2 = 2 \log_{10} 10 = 2 \times 1 = 2$ だからです．

➡ The common logarithm of 100 is 2, because
$100 = 10^2 = 2\log_{10}10 = 2 \times 1 = 2$

  読み方: 100 equals 10 squared being equal to 2 times the logarithm of 10 being equal to 2 times 1 and also equals 2.

○ $y = a^x$ とすれば，x は基数を a とする y の対数です ($x = \log_a y$ と書きます)．

➡ If $y = a^x$ (y equals a to the xth power), then x is the logarithm of y to the base a [written as $x = \log_a y$ (x equals the logarithm of y to the base a)].

○ 基数 10 の対数を，常用対数と呼びます．

➡ Base-10 logarithms are called common logarithms.

○ $657.3\ (= 6.573 \times 10^2)$ の対数を求めなさい．
この対数は $\log 6.573 + 2\log 10$ で，$2 + \log 6.573$，さらに 2.81778 となります．ここで 2 は指標で，0.8178 は仮数です．

➡ Find the logarithm of $657.3\ (= 6.573 \times 10^2)$.

The logarithm of this is $\log 6.573 + 2 \log 10$, which is $2 + \log 6.573$ or 2.81778. Here, 2 is the characteristic and 0.8178 the mantissa.

根号   radical

根号とは $\sqrt{}$ (radical, または, radical sign) のことで, 根号を含む式から, 乗根 (root) が求められます.

例えば, 式 (expression) $\sqrt[3]{64}$ で, 3 を指数 (index), 64 を被開数 (radicand), $\sqrt{}$ を根号 (radical sign) と呼びます. $\sqrt[3]{64} = \sqrt[3]{(4)^3} = 4$ となりますから, この場合, 4 が 64 の 3 乗根です.

根号の基本的な読み方は下記のとおりです.

- $\sqrt[2]{2} = \sqrt{2}{}^{*}$: 2 の 2 乗根 [the (positive または, plus) square root of 2]
- $-\sqrt[3]{3}$: 3 のマイナス 3 乗根 [the negative (または, minus) cube root of 3]
- $\sqrt[4]{-5}$: -5 の 4 乗根 [the (positive または, plus) fourth root of -5]
- $\sqrt[n]{6}$: 6 の n 乗根 [the (positive または, plus) nth root of 6]
- $\sqrt[n]{x^n}$: x の n 乗の n 乗根 [the (positive または, plus) nth root of n to the nth power]

* 2 乗根の場合, 2 は通常, 省略できます.

次の法則にしたがって，根号は簡易化されます．

ただし，ここで使用されている x および y は非負実数 (nonnegative real numbers) です．

$\sqrt[n]{x^n} = x$

x^n の n 乗根は x です．

(The nth root of x to the nth power is x.)

(例)
$$\sqrt{4} = \sqrt{2^2} = 2$$
$$\sqrt[3]{8} = \sqrt[3]{2^3} = 2$$
$$\sqrt[n]{x \cdot y} = \sqrt[n]{x} \cdot \sqrt[n]{y}$$

積の n 乗根はそれぞれの n 乗根の積です．

(The nth root of a product is the product of the nth roots.)

(例)
$$\sqrt{30} = \sqrt{3 \times 10} = \sqrt{3 \times 2 \times 5} = \sqrt{3} \times \sqrt{2} \times \sqrt{5}$$
$$\sqrt{75x^2y} = \sqrt{5 \times 5 \times 3 \times x^2 \times y} = \sqrt{5^2} \times \sqrt{3} \times \sqrt{x^2} \times \sqrt{y}$$
$$= 5 \times \sqrt{3} \times x \times \sqrt{y} = 5x\sqrt{3}\sqrt{y} = 5x\sqrt{3y}$$
$$\sqrt[n]{\frac{x}{y}} = \frac{\sqrt[n]{x}}{\sqrt[n]{y}}$$

分数の n 乗根は，分子の n 乗根を分母の n 乗根で割った数に等しくなります．

(The nth root of a fraction equals the nth root of the numerator divided by the nth root of the denominator.)

(例)
$$\sqrt{\frac{8}{18}} = \sqrt{\frac{2 \times 4}{2 \times 9}} = \sqrt{\frac{4}{9}} = \sqrt{\frac{2^2}{3^2}} = \frac{2}{3}$$

$$\sqrt{\frac{50x^3}{9}} = \sqrt{\frac{5 \times 5 \times 2 \times x^2 \times x}{3^2}} = \frac{\sqrt{5^2} \times \sqrt{2} \times \sqrt{x^2} \times \sqrt{x}}{\sqrt{3^2}}$$

$$= \frac{5 \times \sqrt{2} \times x \times \sqrt{x}}{3} = \frac{5x\sqrt{2}\sqrt{x}}{3} = \frac{5x\sqrt{2x}}{3}$$

注1) 代表的数値の2乗根は下記のとおりです.

$\sqrt{2} \approx 1.414$　　　$\sqrt{3} \approx 1.732$

$\sqrt{5} \approx 2.236$　　　$\sqrt{7} \approx 2.645$

構文例および読み方

● $\sqrt{48x^2}$

→ The square root of 48 times x squared.

● $\sqrt[4]{\frac{16}{81}} = \frac{2}{3}$

→ The fourth root of 16 divided by 81 equals 2 divided by 3.

● $-\sqrt[3]{1} = -1$

→ Minus the cube root of 1 equals minus 1.

● $\frac{5x\sqrt{2x}}{3}$

→ $5x$ times the square root of $2x$ divided by 3.

● $4\sqrt{7} - 6\sqrt{7} = -2\sqrt{7}$

➡ 4 times the square root of 7 minus 6 times the square root of 7 equals minus 2 times the square root of 7.

⊃ $\sqrt{2}$を開平しなさい．

➡ Extract the square root of 2.

⊃ $\sqrt{7}-1$

➡ The square root of 7, the radicand minus 1.

⊃ もし存在する場合は各乗根を求めなさい．

$$\sqrt[3]{64},\ 4\sqrt{16}$$

➡ Find each root if it exists.

$$\sqrt[3]{64},\ 4\sqrt{16}$$

The cube root of 64, and 4 times the square root of 16.

⊃ 625 の 4 乗根は 5 です（すなわち$\sqrt[4]{625}=5$）．

➡ The fourth root of 625 equals 5 (i.e. $\sqrt[4]{625}=5$).

分数　　　　　　　　fraction，または，fractional number

1つの数 a ［または，式 (expression)］を，別の数 b（または，式）で割って得られた商 (a quotient) c で，

$$c = \frac{a}{b}$$

と表されます．

　被除数 a (dividend) を分数の分子 (numerator)，除数 b (divisor) を分母 (denominator) と呼びます．

　分数を既約分数 (the lowest term) に通分する (to reduce) には，分子と分母を因数分解して (to factor) から，共通因数 (common factor) を消去します (to cancel)．

(例)　$\dfrac{4x^3}{8x^2} = \dfrac{1}{2}x$

　分数は一般に次のように分類されます．

- common (または，simple, vulgar) fraction：単分数
  分子，分母ともに整数の場合をいいます．

(例)　$\dfrac{1}{2}$

- complex fraction：繁分数または重分数，複分数
  分子，分母自体が分数の場合をいいます．

(例)　$\dfrac{\frac{1}{2}}{\frac{3}{4}}$

- proper fraction：真分数
  分子が分母より小さい場合をいいます．

(例)　$\dfrac{7}{8}$

- improper fraction：仮分数

分子が分母より大きい場合をいいます．

(例)　$\frac{8}{7}$

- mixed fraction：　帯分数
  真分数と整数 (integer) からなる場合をいいます．

(例)　$1\frac{1}{2}$

　分数は，整数 (integer) と小数 (decimal fraction) の組合せでも表現されます．整数と小数は小数点 (a decimal point) で分離されます．

(例)　$1\frac{1}{2}=1.5,\ \frac{1}{2}=0.5$

構文例および読み方

○　$\frac{1}{4} \cdot \frac{3}{8} = \frac{3}{32}$

➡ One-fourth times three-eighths equals three thirty-seconds.

○　$\frac{2}{5} \div \frac{4}{25} = \frac{5}{2} = 2\frac{1}{2}$

➡ Two-fifths divided by four twenty-fifths equals five-seconds and also equals two and one-half.

○　$\frac{6x}{11} \cdot \frac{2}{5y} = \frac{12x}{55y}$

→ 6x over 11 times 2 over 5y equals 12x over 55y.

○ $\dfrac{3x^2}{5} \div \dfrac{2x}{y} = \dfrac{3xy}{10}$

→ 3 times x squared over 5 divided by 2x over y equals 3xy over 10.

○ $\dfrac{2}{y^2} - \dfrac{3}{y} = \dfrac{2-3y}{y^2}$

→ 2 divided y squared minus 3 divided by y equals 2 minus 3y divided y squared.

○ 重分数は，分子 (および/または分母) に分数を含む有理式です．

→ A complex fraction is a rational expression that contains fractions in the numerator and/or denominator.

○ 小数は一連の分数です．

→ A decimal fraction is a series of fractions.

○ 分数 25/100 は小数では 0.25 と書かれます．

→ The fraction 25/100 is written 0.25 as a decimal.

○ 1/3 および 5/8 は分数です．

→ 1/3 and 5/8 are fractions.

注) x/y を x divided by y とするか x-yths (序数表示) とするかは, x と y の複雑さで決めます.

小数   `decimal,または,decimal fraction`

基数10にもとづいた計数システムにより表された, 真分数のことです (proper fraction expressed by a system of counting based on the base 10).

すなわち, 1より小さい数のことです.

(例)

$$0.537 = \left(0 + \frac{5}{10} + \frac{3}{100} + \frac{7}{1000} \right)$$
$$= 0 + (5 \times 10^{-1}) + (3 \times 10^{-2}) + (7 \times 10^{-3})$$
$$= 0 + 0.5 + 0.03 + 0.007$$

また, 帯小数 (mixed decimal) は整数 (integer) と小数 (decimal fraction) より構成されます.

(例) $3127.635 = 3127 + 0.635$

上記の例で, 整数3127は整数部 (the integer part) に, 小数0.635は小数部 (the fraction part) に含まれ, 各部は1つの点 (a dot) で分離されています. この点を小数点 (a decimal point) と呼びます.

それぞれの数の位置関係は下記のとおりです.

3127.635

- $\dfrac{1}{1000}$ の位または小数点以下第 3 位 (thousandths, または, the third decimal place)
- $\dfrac{1}{100}$ の位または小数点以下第 2 位 (hundredths, または, the second decimal place)
- $\dfrac{1}{10}$ の位または小数点以下第 1 位 (tenths, または, the first decimal place)
- decimal point
- 1 の位 (ones place)
- 10 の位 (tens place)
- 100 の位 (hundreds place)
- 1000 の位 (thousands place)

3127.6 → 小数点以下 1 桁 (one decimal place)
3127.63 → 小数点以下 2 桁 (two decimal places)
3127.635 → 小数点以下 3 桁 (three decimal places)

以上から,

・小数点右側の最初の数字は 6 です.

The first figure (または, digit, numeral) to the right of the decimal point is 6.

・$\dfrac{1}{10}$ の位の数字は 6 です.

The figure (または, digit, numeral) in the tenths place is 6.

・小数点の次,最初の数字は6です.

The first figure (または, digit, numeral) following the decimal point is 6.

・小数点以下第1位の数字は6です.

The figure (または, digit, numeral) in the first decimal place is 6.

・小数点右側の最初の数字は6です.

The figure (または, digit, numeral) of the first position to the right of the decimal point is 6.

となり,上記の5文はいずれも同じ意味です.

・小数は
$$\frac{1}{10}$$
の位の数,
$$\frac{1}{100}$$
の位の数,
$$\frac{1}{1000}$$
の位の数などで構成されます.

A decimal fraction consists of a number of tenths plus a number of hundredths plus a number of thousandths, etc.

同様に,

・小数点左側の 2 番目の数字は 2 です．

The second figure (または，digit, numeral) to the left of the decimal point is 2.

・10 の位の数字は 2 です．

The figure (または，digit, numeral) in the tens place is 2.

となり，上記の 2 文はいずれも同じ意味です．

・1 の位の数字は 7 です．

The figure (または，digit, numeral) in the ones place is 7.

・100 の位，10 の位および 1 の位の情報は大型サイズの数字で表示され，小数点第 1 位の情報は小型サイズの数字で表示されます．

The hundreds, tens, and ones information will be displayed by the large digits and the tenths information will be displayed by the small digits.

小数には，次の種類があります．

(i) finite (または，terminating) decimal：有限小数
　　小数点以下の桁数が有限な小数．

(ii) infinite (または，nonterminating) decimal：無限小数
　　小数点以下の桁数が無限な小数．
無限小数には，さらに次のものが含まれます．
(ii-1) repeating (または，recurring) decimal：循環小数
　　小数点以下第 1 位の数字が無限にくり返される小数．

(例)

$0.333\cdots = 0.\dot{3}$*

* 3の上部に打たれた "˙" は，その下の3が無限にくり返される [zero (または, naught) point three recurring] ことを示しています．

(ii-2) periodic decimal： 周期小数

小数点以下，ある桁数の数字が無限にくり返される小数．

(例)

$0.714\,285\,714\,285\cdots = 0.\dot{7}14\,28\dot{5}$

(iii) nonrepeating (または, nonperiodic) decimal： 非循環小数

小数点以下の数字が無限で，くり返される部分のない小数．

(例)

$\pi = 3.141\cdots$

$e = 2.718\cdots$

構文例および読みかた

● $\dfrac{3}{5} = 0.6$

➡ Three-fifths equals zero point 6.

● $\dfrac{x^2-1}{x-1}$

➡ *x* squared minus 1 over *x* minus 1.

3　四則演算
four basic operations

- $\dfrac{3.6 \times 10^{-2}}{6 \times 10^{3}} = 0.6 \times 10^{-5}$

→ 3 point 6 times 10 to the minus second power divided by 6 times 10 to the third power equals zero point 6 times 10 to the minus fifth power.

- $23.6 + 1.75 + 1.023 = 26.373$

→ Twenty three point six plus one point seven five plus one point zero two three equals twenty six point three seven three.

- $4\dfrac{3}{4} - 1\dfrac{1}{2} = 3\dfrac{1}{4} = 3.25$

→ Four and three-quarters minus one and one-half equals three and one-quarter and also equals three point two five.

- $\dfrac{1}{3} = 0.\dot{3}$

→ One-third equals zero point three recurring.

- $\dfrac{5}{7} = 0.\dot{7}14\,28\dot{5}$

→ Five-sevenths equals zero point seven one four two eight five with the block of digits from seven to five repeated endlessly.

- 483 の 10 の位の数字は何ですか？　8 です．

➡ Which digit is the tens place in 483 ? The digit in the tens place is 8.

⊃ 45,328.769 のどの位に,数字 7 がありますか? 小数点以下第 1 位です.

➡ The digit 7 is in which place in 45,328.769 ? It is in the tenths place.

四捨五入する,丸める   rounding off

To change (an exact figure) into the nearest whole number.

ある小数点以下 (after a certain decimal place) の数字 (figures,または,digits) を切り捨てる[to drop (または,omit, ignore, discard)] ことで,小数を近似する (to approximate) 一手法のことです.

(i) 切り捨てられる最初の数字が,5 より大きい,または 5 に等しい[the first digit (または,figure) dropped is greater than or equal to 5] 場合は,その直前の数字 [the preceding digit (または,figure)] に 1 が加算されます.

(ii) 切り捨てられる最初の数字が,5 より小さい (the first digit dropped is less than 5) 場合,その直前の数字は変化しません.

例えば,

1.575 3 is rounded off to 1.58 to two decimal places.

上文を直訳すると,「1.5753 を小数点以下 2 桁に四捨五入*すると,1.58 が得られます」となりますが,次のように訳すとさらに意味が明確になります.

「1.5753 の小数点第 3 位以下を四捨五入すると,1.58 が得られます.」

* 1.5753 を小数点 2 桁に四捨五入するということは,得られた数字を小数点以下 2 桁の数 1.58 にすることです.すなわち,小数点第 3 位およびそれ以下の数字を四捨五入して,小数点以下 2 桁の数字 1.58 を求めることと同じ意味です.

構文例および読み方

○ この式を小数点以下第 3 位まで計算しなさい.

→ Calculate this equation down to the third decimal place.

○ 1.5742 の小数点第 3 位以下を四捨五入すると,1.57 が得られます.

→ 1.5742 is rounded off to 1.57 to two decimal places.

○ 1.5753 の小数点第 3 位以下を四捨五入しなさい.

→ Round off 1.575 3 to two decimal places.

⇨ 小数点第1位以下の数字を四捨五入しなさい．

➡ Round off the digits (または，figures) after one decimal place.

⇨ 9.45の小数点第2位以下を四捨五入すると，9.5となります．

➡ 9.45 is rounded off to 9.5 to one decimal place.

⇨ 次の式を計算し，求められた値の小数点第3位以下を四捨五入しなさい．

➡ Calculate the following equation and then round off the value thus obtained to two decimal places.

● 端数の切り捨て (truncating)

(i) 小数点以下の端数を切り捨てること．
(Dropping digits of a number that are to the right of the decimal point.)
　この場合は，小数点以下の数字をそれらの大小に関係なく，すべて切り捨てて整数とすることを意味します．
(例)
　　5.585 → 5
　　737.984 → 737

構文例

⇨ 6.45の小数点以下の端数を切り捨てると6が得られます．

また，737.984 の小数点以下の端数を切り捨てると 737 が得られます．

➡ Truncating 6.45 gives 6, and truncating 737.984 gives 737.

◐ 小数点以下の端数を切り捨てなさい．

➡ Truncate the digits (または，figures) below the decimal point.

(ii) 小数点第何位以下の端数を四捨五入することなく，切り捨てること．
(The process of approximating a decimal by dropping digits after a certain decimal place without rounding off.)

構文例

◐ 1.5753 の小数点第 3 位以下の端数を切り捨てると，1.57 が得られます．

➡ 1.5753 is truncated to 1.57 to two decimal places.

◐ 232.534 5 の小数点第 4 位以下の端数を切り捨てなさい．

➡ Truncate 232.534 5 to three decimal places.

注) 以上から，例えば単に「232.33 の端数を切り捨てなさい」(Truncate 232.33) といえば，232 とすることを意味し，「232.33 の小数点第 2 位以下の端数を切り捨てなさい」(Truncate 232.33 to one decimal place) といえば 232.3 とする

ことを意味します．

　すなわち，2通りの使われ方があるので注意が必要です．

・rounding down：数，価格などの小数点以下の端数を切り捨てて，それに最も近い整数値とする．

To decrease a number, price, etc. to the nearest whole number

意味上からは truncating の (i) と同じです．
例えば，

・月収が 12,386.46 円の場合は，税の目的から小数点以下の端数を切り捨てて 12,386 円とされます．

If your income is ¥12,386.46 a month, it will be rounded down to ¥12,386 for tax purposes.

rounding up：rounding down と対照的に数，価格などの小数点以下の端数を四捨五入して切り上げ，それに最も近い整数値とする．

To increase a number, price etc. to the nearest whole number

例えば，

・数字 301.522 は，この計算を行うために，小数点以下の端数を切り上げて 302 とされます．

The figure 301.522 is rounded up to 302 for this calculation.

比率，割合，〜の比   ratio

A figure showing the number of times one quantity contains another, used to show the relationship between two amounts.

2つの数，または数量の商のことです．2つの数，または数量を a および b として，その比率を

a/b, $\dfrac{a}{b}$, $a:b$, または, a 対 b (a to b)

と書きます．読み方はいずれの場合も a to b です．

この場合，a を前項 (antecedent)，b を後項 (consequent) と呼びます．

構文例および読み方

○ $a:b=c:d$

➡ The ratio of a to b is c to d.

○ このクラスの男子と女子の比率は $3:1$ です．

➡ The ratio of boys to girls in this class is $3:1$ (または, three to one).

○ (エネルギー出力) 対 (エネルギー入力) の比率は，パーセントで表されます．

➡ The ratio of energy output to energy input is expressed as a percentage.

- ◯ (負荷) 対 (作動力比) を機械的利益と呼びます.

- ➡ The ratio of load/effort is called the mechanical advantage.

- ◯ 10 対 5 の比率は 2 対 1 です.

- ➡ The ratio of 10 to 5 is 2 to 1.

- ◯ このエンジンの圧縮比は 8 対 1 です.

- ➡ The engine has a compression ratio of $8:1$.

- ◯ このような人々は 1 対 1000 の割合で存在します.

- ➡ These people exist in a (または, the) ratio of 1 to 1000.

比例，比例算　　　　　　　　　　　　`proportion`

比例算は，2 つの比率が互いに等しい，として書かれます．いま $a:b$ が $c:d$ に等しい場合，下記のように書くことができます．

$$a:b=c:d, \text{ または, } \frac{a}{b}=\frac{c}{d}$$

$a/b = c/d$ なら $ad = bc$ が成り立ちます．

構文例および読み方

➲ 次の比例算を解きなさい．

$$\frac{x}{6} = \frac{8}{4}$$

上式から，
$$4x = 48,$$
ゆえに，
$$x = \frac{48}{4} = 12$$

➡ Solve the following proportion.

$$\frac{x}{6} = \frac{8}{4}$$

読み方: x over 6 equals 8 over 4.

From the above,
$$4x = 48$$
読み方: $4x$ equals 48.

Therefore,
$$x = 12$$
読み方: x equals 12.

➲ 40 は 200 の何％ですか？

$$\frac{x}{100} = \frac{40}{200}$$

上式から，$200x = 4000$

$x = 20\%$

➡ 40 is what percent of 200 ?

$$\frac{x}{100} = \frac{40}{200}$$

読み方: x over 100 equals 40 over 200.

From the above,
   $200x = 4000$
Therefore, $x = 20\%$

◆ 30の15％に相当する数は？

$$\frac{15}{100} = \frac{x}{30}$$

上式から，
   $100x = 450$
ゆえに，
   $x = 4.5$

➡ What number is 15 % of 30 ?
$$\frac{15}{100} = \frac{x}{30}$$
From the above, $100x = 450$
Therefore, $x = 4.5$

3　四則演算
four basic operations

4 線 lines

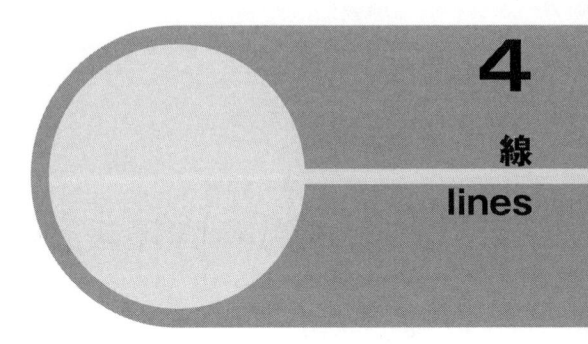

直線　　　　　　　　　　　　　　　　　　straight line

直線は 2 点間の最短距離として表されます．

$$\xleftarrow{\quad\overset{A}{\bullet}\qquad\qquad\overset{B}{\bullet}\quad}\rightarrow$$

線 AB (line AB)

線 AB は，$\overleftrightarrow{AB}$ (the symbol $\leftrightarrow$ written on top of the two letters A and B) と書きます．

線分　　　　　　　　　　　　　　　　　　line segment

線分とは，1 本の線の一部分 (a piece of a line) のことです．線分は 2 つの端点 (two endpoints) をもっています．

$$\xleftarrow{\quad\overset{A}{\bullet}\quad\overset{C}{\bullet}\;\overset{D}{\bullet}\quad\overset{B}{\bullet}\quad}\rightarrow$$

線分 CD (line segment CD)

線分 CD は，$\overline{CD}$ (the symbol $-$ written on top of the two letters C and D) または CD と表します.

線分 $\overline{CD}$ は線 $\overleftrightarrow{AB}$ の一部分です.

半直線　　　　　　　　　　　　　　　ray, または, half line

半直線は 1 つのみの端点 (an endpoint), または, 原点 (the origin) をもち, 1 方向に連続しています.

```
A       B            C       B
●───────●──▶    ◀────●───────●
```

半直線 AB

半直線 AB は $\overrightarrow{AB}$ と表します.

交差線　　　　　　　　　　　　　　　intersecting lines

1 点で交わる 2 本, または, それ以上の線を交差線と呼びます.

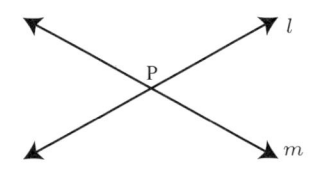

上図では 2 本の線 l, m は P で交わります (to intersect at P).

垂線　　　　　　　　　　　　　　　perpendicular lines

直角 (right angles : 90°) をなすように交わる 2 本の線を, 垂線と呼びます.

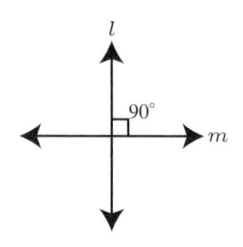

記号⊥は垂線を表します (The symbol ⊥ is used to denote perpendicular lines.).

すなわち，上図では line l ⊥ line m と書かれます.

読み方: The line l is perpendicular to the line m.　または, The lines l and m meet at right angles.

記号∟は直角を表します.

平行線　　　　　　　　　　　　　　　　　　　`parallel lines`

常に同一間隔を保つ2本，またはそれ以上の線を平行線と呼びます.

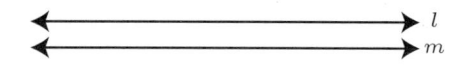

記号∥は平行線を表します.

すなわち，上図では line l ∥ line m と書きます.

読み方: The line l is parallel to the line m.

横断線で切られる平行線　parallel lines cut by transversal

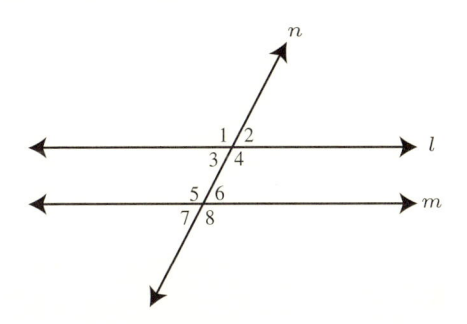

図で線 n が横断線 (transversal) で，線 l と m は平行線です．
これにより，8つの角度 (8 angles) が形成されます．

記号 ∠ は角度を表します．図の1の角度は ∠1 (angle 1)，8の角度は ∠8 (angle 8) と書きます．

これらの角度をそれぞれ，次のように呼びます．

- 隣接角 (adjacent angles)
    ∠1 と ∠2, ∠2 と ∠4, ∠3 と ∠4, ∠1 と ∠3,
    ∠5 と ∠6, ∠7 と ∠8, ∠5 と ∠7, ∠6 と ∠8
これらの2つの角を加算すると $180°$ となります．
(例)
    $∠1+∠2=180°$
    読み方：The angle 1 plus the angle 2 equals 180 degrees.

- 頂角 (vertical angles)
    ∠1 と ∠4, ∠2 と ∠3, ∠5 と ∠8, ∠6 と ∠7

対向頂角どうしは等しい角度です．
(例)

∠2=∠3

読み方: The angle 2 equals the angle 3.

● 同位角 (corresponding angles)

線 l と線 m を重ね合わせた場合，互いに一致して等しくなる角度のことです (the angles that coincide with each other).

∠1と∠5, ∠3と∠7, ∠2と∠6, ∠4と∠8

(例)

∠1=∠5

● 錯内角 (alternate interior angles)

錯角 (alternate angles) は，横断線の対向側に (on the opposite side of the transversal) 形成されます．

内角 (interior angles) は平行線の内側の角度のことです．
外角 (exterior angles) は平行線の外側の角度のことです．

したがって，錯内角 (alternate interior angles) は∠3と∠6, ∠4と∠5で，∠3=∠6, ∠4=∠5となります．

錯外角 (alternate exterior angles) は，∠2と∠7, ∠1と∠8で，∠2=∠7, ∠1=∠8となります．

● 連続内角 (consecutive interior angles)

連続内角は，横断線の同じ側 (on the same side of the transversal) の角度のことです．

したがって，連続内角は∠3と∠5, ∠4と∠6で，それぞ

れ加算して 180°となります.

構文例および読みかた

○ ∠2+∠4=180°

➡ The angle 2 plus the angle 4 equals 180 degrees.

○ 2本の線は直角に交わります.

➡ Two lines cross (または, meet) at right angles.

○ 線 AB は線 CD と平行です.

➡ The line AB is parallel with the line CD.

○ 線 AB に垂直な 1 本の線を引きなさい.

➡ Draw a line perpendicular(または, at right angles) to the line AB.

○ 鋭角は, 0°より大きく 90°以下の角度です.

➡ Acute angle is greater than 0 degrees but less than 90 degrees.

○ 鈍角は, 90°より大きく 360°以下の角度です.

➡ Obtuse angle is greater than 90 degrees but less than 360 degrees.

5 統計 statistics

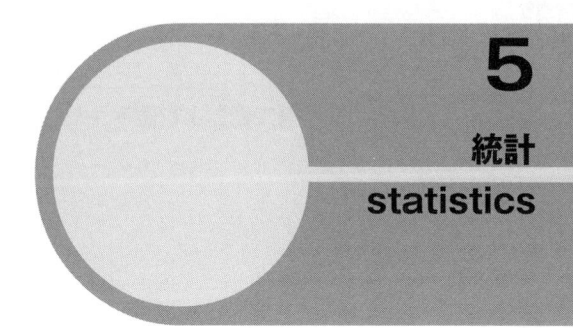

平均値，または，算術平均値
the mean(arithmetic mean)または,the average

平均値を得るためには，すべての項目の合計値を求めて，それを項目数で割ります．

構文例および読み方

◯ 25, 27, 27 および 27 の平均値はいくつですか？

$$25+27+27+27=106$$
$$106 \div 4 = 26\frac{1}{2}$$

平均値は

$$26\frac{1}{2}$$

です．

➡ What is the mean of 25, 27, 27, and 27 ?

25 plus 27 plus 27 plus 27 equals 106.

106 divided by 4 equals 26 and one-half.
The mean is 26 and one-half.

◯ 20 と −10 の平均値はいくつですか？

$$20+(-10)=10$$
$$10\div 2=5$$
平均値は 5 です．

➡ What is the mean of 20 and −10 ?

20 plus the additive inverse of 10 equals 10.
10 divided by 2 equals 5.
The mean value is 5.

◯ 3, 7, −5, −13 の平均値を求めなさい．

$$3+7+(-5)+(-13)=-8$$
$$-8\div 4=-2$$
平均値は −2 です．

➡ Find the mean of 3, 7, −5, and −13.
3 plus 7 plus the additive inverse of 5 plus the additive inverse of 13 equals minus 8.
Minus 8 divided by 4 equals minus 2.
The mean is minus 2.

重みつき平均値　　　　　　　　weighted mean

1つ (または，それ以上) の項目が，そのほかの項目より

多く使用されるとき，より重み (weight) をもっているといいます．

例えば，3回のテストで得られた平均値70点と7回のテストで得られた平均値85点を考えてみます．テスト回数の合計は10回です．

この場合の重みつき平均値は下記より得られます．

$3 \times 70 = 210$

$7 \times 85 = 595$

$(210 + 595) \div 10 = 805 \div 10 = 80.5$

よって，重みつき平均値は80.5点となります．

構文例および読み方

● あるテストで6名の学生の平均値は90点，同じテストでほか4名の学生の平均値は70点です．全学生の平均点は？

$6 \times 90 = 540$

$4 \times 70 = 280$

$(540 + 280) \div 10 = 820 \div 10 = 82$ 点

➡ Six students averaged a score of 90 on a test. Four other students averaged a score of 70 on the test. What was the mean score of all ten students ?

6 times 90 equals 540.
4 times 70 equals 280.
540 plus 280 in the parentheses divided by 10 equals 82. The mean score is 82.

中央値   median

増加, または減少方向順に配列された1組の数の中央値は, この1組の数の合計が奇数の場合はちょうどまん中の数 (the middle number), 一方, 数の合計が偶数の場合はまん中の2つの数 (the middle two numbers) の平均値にあたります.

構文例および読み方

● 次の数のグループにおける中央値を求めなさい.

　　9　3　5
数を増加方向順に並べ替えると,
　　3　**5**　9
したがって, 中央値は5です.

➡ Find the median of the following group of numbers.
　　9　3　5
If the above group of numbers is re-arranged in ascending order, the following results.
　　3　5　9
Therefore, the median is 5.

● 次の数のグループにおける中央値を求めなさい.

18, 16, 0, 7, 12　→　0, 7, **12**, 16, 18
したがって, 中央値は12です.

➡ Find the median of the following group of numbers.

5　統計
statistics

18, 16, 0, 7 and 12 become 0, 7, **12**, 16 and 18, if re-arranged in ascending order. Therefore, the median is 12.

最頻値　　　　　　　　　　　　　　　mode,または,modal class

1組の数，または集合数で，その現れる頻度が最も大きい数を最頻値といいます．

構文例および読み方

◆ 3, 4, 8, 9, 9, 2, 6, 11 の最頻値を求めなさい．

最頻値は9です．というのは，他の数より多くの頻度で現れているからです．

➡ Find the mode of 3, 4, 8, 9, 9, 2, 6, 11.

The mode is 9, because it appears more often than any other numbers.

◆ 7, 8, 4, -3, 2, -3 の最頻値を求めなさい．最頻値は-3です

➡ Find the mode of 7, 8, 4, -3, 2, -3. The mode is -3.

6 多角形
polygons

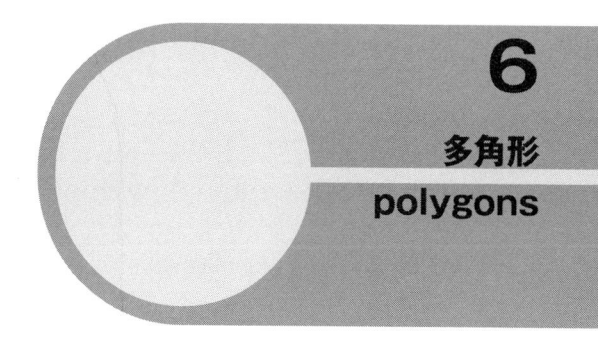

　3つ，またはそれ以上の辺 (three or more sides) をもつ平面 (plane) 上の閉じた形，または図を多角形と呼びます．

三角形　　　　　　　　　　　　　　　　　　　　　triangle

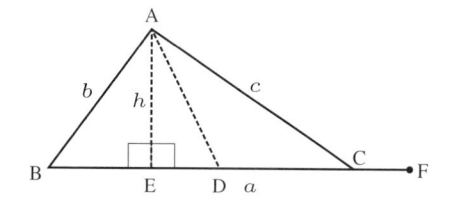

　三角形はその内側に3つの角度をもっており，これら角度の合計は常に $180°$ ($\angle A + \angle B + \angle C = 180°$) となります．
　$\angle A$, $\angle B$, $\angle C$ を内角 (interior angles) と呼びます．

　三角形は記号 $\triangle$ で表され，上記を $\triangle ABC$ と書きます．BC (bottom side) を底辺 (base) と呼び，頂点 (vertex) A から底辺 (base) に引いた垂線 (perpendicular line) AE を高さ，(height，または，altitude) と呼びます．この場合，$AE \perp BC$ となります．

さらに，D を BC の中点 (midpoint) (BD＝DC) とし，A から D に線を引いた場合，AD を △ABC の中線 (median) と呼びます．

底辺 BC を F 点まで延長した場合，外角 (∠ACF) (exterior angle) が形成されます．

● 面積 (the area of a triangle)

三角形の面積は下記の公式で表されます．

$A = \frac{1}{2}ah$

読み方：The area of a triangle equals one-half times the length of the base a times the height h.

● 辺による三角形の分類

・正三角形 (equilateral triangle，または，equiangular triangle)
 3 辺の長さが等しい三角形のことです．

・2 等辺三角形 (isosceles triangle)
 2 辺の長さが等しい三角形のことです．

・不等辺三角形 (scalene triangle)
 各辺の長さが等しくない三角形のことです．

● 角度による三角形の分類

・直角三角形 (right triangle)
 内角の 1 つが 90°の三角形のことです．

直角三角形の場合，各辺 (each side) の長さ(BC＝a，AB＝

b, AC $=c$) の間にはピタゴラスの定理 (the Pythagorean theorem, または, Pythagoras' theorem) が成立します.

$$a^2+b^2=c^2$$

読み方: a squared plus b squared equals c squared.

・鈍角三角形 (obtuse triangle)

内角の1つが鈍角 (obtuse angle: greater than 90° but less than 180°) の三角形のことです.

・鋭角三角形 (acute triangle)

内角のすべてが鋭角 (acute angles: greater than 0° but less than 90°) の三角形のことです.

構文例および読み方

● △ABC では,
AB+BC＞AC
AB+AC＞BC
AC+BC＞AB.

➡ In a triangle △ABC, AB plus BC is longer than AC; AB plus AC is longer than BC; and AC plus BC is longer than AB.

● 直角三角形で辺 AB の長さを 5, BC の長さを 7, AC の長さを c とする. この直角三角形の c の長さを求めなさい.

ピタゴラスの定理から,
$$5^2+7^2=c^2$$
$$74=c^2$$

6 多角形
polygons

$c = \sqrt{74}$

➡ Let 5 be the length of the side AB, 7 be the length of the side BC and c be the length of the side AC of a right triangle. Find c in this right triangle.

The following equation is obtained using the Pythagorean theorem.

$5^2 + 7^2 = c^2$
読み方: 5 squared plus 7 squared equals c squared.

$74 = c^2$
読み方: 74 equals c squared.

$c = \sqrt{74}$
読み方: c equals the square root of 74.

● 三角形の2つの角度は45°と85°です．残りの角度は何度ですか？

三角形の3つの角度の合計は180°です．45°と85°の合計は130°となります．それゆえ，残りの角度は180°−130° = 50°です．

➡ Two angles of a triangle measure 45 degrees and 85 degrees. How many degrees are there in the remaining angle ?

The sum of three angles equals 180 degrees. The sum of 45 degrees and 85 degrees is 130 degrees. Therefore, the remaining angle is 180 degrees minus 130 degrees and also equals 50 degrees.

❯ △ABC の底辺を 5 cm，高さを 3 cm とした場合の面積 A を求めなさい．

$$A = \frac{1}{2} \cdot 5 \cdot 3 = \frac{15}{2} = 7.5 \text{cm}^2$$

➡ Let 5 cm be the length of the base and 3 cm be the length of the height of a triangle. Find the area of this triangle.

$$A = \frac{1}{2} \cdot 5 \cdot 3$$

読み方： A equals one-half times 5 times 3.
Therefore, the area of this triangle is 7.5 square centimeters.

四辺形　　　　　　　　　　　　　　　quadrilaterals

　　4つの辺 (four sides) をもつ多角形 (polygon) を四辺形と呼びます．

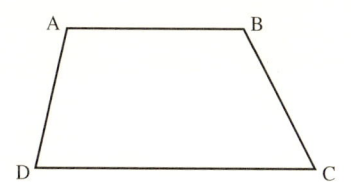

　　内側には4つの角度，すなわち内角 (interior angles) があります．これら4つの角度の合計 (∠A+∠B+∠C+∠D) は常に 360° となります．

　　四辺形は ABCD と表します．

6　多角形
polygons

● 四辺形の種類

・正方形 (square)

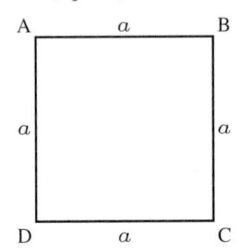

```
    A      a      B
    
    a             a
    
    D      a      C
```

正方形は4つの等しい辺と，4つの直角をもっています．

対向辺 (opposite sides) は平行です．正方形の対角線 (diagonals) の長さは等しく，お互いを2等分 (to bisect) します．

・矩形 (rectangle)

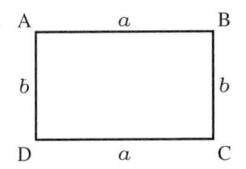

```
    A      a      B
    
    b             b
    
    D      a      C
```

矩形は互いの対向辺の長さが等しく，平行で4つの直角をもちます．矩形の対角線の長さは等しく，お互いを2等分します．

・平行四辺形 (parallelogram)

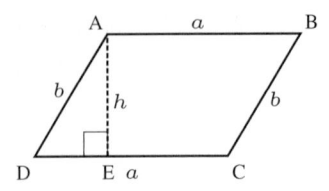

```
        A      a      B
       /       |      /
      b        h     b
     /         |    /
    D     E  a    C
```

平行四辺形の対向辺の長さは互いに等しく，しかも，平行で対向角度 (opposite angles) も等しくなります．

∠A=∠C, ∠B=∠D
∠A+∠B=180°
∠A+∠D=180°
∠B+∠C=180°
∠C+∠D=180°
AE : 高さ (height), h
AB∥DC, AD∥BC

・斜方形，または，ひし形 (rhombus，複数形 : rhombuses, または，rhombi)

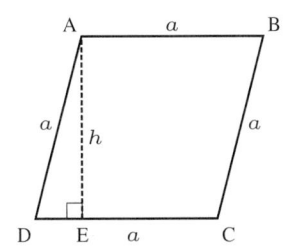

斜方形は4つの等しい長さの辺をもつ平行四辺形です．
AE : 高さ (height), h

・台形 (trapezoid)

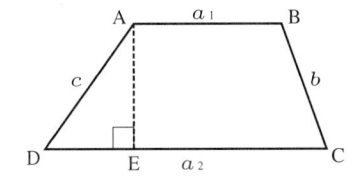

6 多角形
polygons

台形はただ1対の平行辺 (only one pair of parallel sides) をもっています.

上図の台形では，CD および AB が底辺 (bases) で，平行でない辺 (nonparallel sides: AD と BC) は辺 (legs) と呼びます.

● **面積** (the area of a trapezoid)

台形の面積は次の公式で表されます.

$A = \dfrac{1}{2}(a_1 + a_2)h$

ここで，a_1, a_2 は底辺の長さ，h は高さ

読み方: The area of a trapezoid equals one-half on the outside of the parentheses times a_1 plus a_2 on the inside times h, where a_1 and a_2 are the lengths of the bases and h is the height.

面積 A で $a_1 = a_2 = a$, $h = a$ とすると，四角形の面積 ($A = a^2$) です。$a_1 = a_2 = a$, $h = b$ とすると，矩形の面積 ($A = ab$) です．また，$a_1 = a_2 = a$ とすると，平行四辺形の面積 ($A = ah$)．および，$a_1 = a_2 = a$ とすると，斜方形の面積 ($A = ah$) となります．

そのほかの多角形

上記以外の多角形については，名称のみを記載します．

・五角形 (pentagon)
・六角形 (hexagon)
・七角形 (heptagon)

- 八角形 (octagon)
- 九角形 (nonagon)
- 十角形 (decagon)

構文例および読み方

◉ 多角形の内角の合計は，公式 $(n-2)180°$ (n：多角形の辺数) を使用して決定できます．

例えば四角形では $n=4$ なので，
$$(4-2)\,180° = 360°$$
です．

➡ The sum of the interior angles in any polygon can be determined by using the formula :

$$(n-2)\,180°$$
読み方： n minus 2 in the parentheses times 180 degrees

where n is the number of sides in the polygon.

For example, as a quadrilateral has 4 sides, the sum of the interior angles in the quadrilateral is $360°$ from $(4-2)180°$.

◉ 次の台形の周囲 P と面積 A を求めなさい．

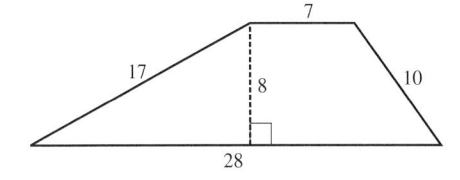

周囲 P は各辺の合計ですから,

$$P=17+7+10+28=62$$

です. 面積は,

$$A=\frac{1}{2}(7+28)\,8=140$$

です.

➡ Find the perimeter P and area A of the following trapezoid.

As the perimeter of the trapezoid is the sum of all sides, P being equal to 17 plus 7 plus 10 plus 28 equals 62.
The area becomes as follows.

$$A=\frac{1}{2}(7+28)\,8=140$$

読み方: A equals one-half on the outside of the parentheses times 7 plus 28 on the inside times 8 and also equals 140.

7

円, 扇形, および楕円
circle, sector, oval

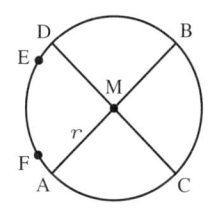

円 circle

平面上で, すべての点が1つの与えられた点より等距離にある場合, 円と呼びます.

上図で,
    M: 中心点 (center point)
    MA=MB: 半径 r (radius)
    AB: 直径 d (diameter)
    AB=2MA=2MB (直径は半径の2倍.)
    EF: 円弧 (arc)

円弧は円周上の2点間距離 ($\overarc{EF}$) です.
記号 "⌒" が円弧を表すために使用されます.

円弧は，度 (または，ラジアン：radian) で測ります．

(例)

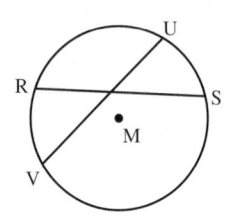

$\overparen{\text{EF}} = 90°$ (または，$\frac{\pi}{2}$)

RS, UV を弦 (chord) と呼びます．
直径は，円で最も長い弦のことです．

● **面積** (the area of a circle)
円の面積は次の公式で表されます．
$A = \pi r^2$

読み方：The area A of a circle equals π(pi) times r squared.

● **円周** (the circumference of a circle)
円の円周は次の公式で表されます．
$C = \pi d = 2\pi r$

読み方：The circumference C of a circle equals π times d, namely, 2 times π times r.

注) ラジアン (radian)
r が円の半径，ℓ がある角度 θ に対する円弧の長さとした場合，その角度は，

$$\theta = \frac{\ell}{r} \; [ラジアン\;(radians)]$$

となります.

1回転の角度 360° は，円周 $2\pi r$ を半径 r で割った値，すなわち 2π ラジアン (2π radians) となります.

$$360° = 2\pi \;[ラジアン]$$
$$180° = \pi \;[ラジアン]$$
$$90° = \frac{\pi}{2} \;[ラジアン]$$

度をラジアンに変換するためには (to convert degrees into radians)，その角度に $\pi/180$ をかけます (to multiply that degree by $\pi/180$).

$\pi = 3.14\cdots$ です.

扇形 sector

2つの半径と，それにより切断されるいずれかの円弧間にある，円の一部のことです.
(例)

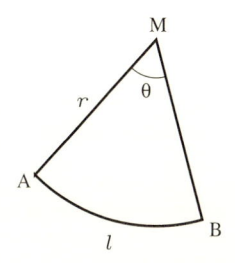

r : 円の半径
l : 円弧 $\overset{\frown}{AB}$ の長さ

● **面積** (the area of a sector)

扇形の面積は，次の公式で表されます．

$$A = \frac{1}{2}r^2\theta$$

ここで，r は半径，θ は2つの半径と円弧間にできるラジアン表示の角度のことです．

読み方: The area A of a sector equals one-half times r squared times θ (theta), where r is the radius and θ is the angle in radians subtended by the arc at the center of the circle.

構文例および読み方

◯ 半径が 3 cm の円の面積と円周を求めなさい．

$A = \pi(3)^2 = 9\pi$ cm^2
$C = 2\pi(3) = 6\pi$ cm

➡ Find the area and circumference of a circle with a radius of 3 cm.

$A = \pi(3)^2 = 9\pi$ cm^2
読み方: A equals π times 3 squared, namely, 9 times π square centimeters.

$C = 2\pi(3) = 6\pi$ cm
読み方: C equals 2times π times 3, and also equals 6 times π centimeters.

◯ r が 3 cm，θ が $90°(\pi/2)$ の扇形の面積を求めなさい．

$$A = \frac{1}{2}(3)^2 \frac{\pi}{2} = \frac{9\pi}{4} \text{ cm}^2$$

➡ Find the area A of a sector, where r is 3cm and θ is $90°$ $(\pi/2)$.

$$A = \frac{1}{2}(3)^2 \frac{\pi}{2} = \frac{9\pi}{4} \text{ cm}^2$$

読み方: A equals one-half times 3 squared times π over 2, namely, 9 times π over 4 square centimeters.

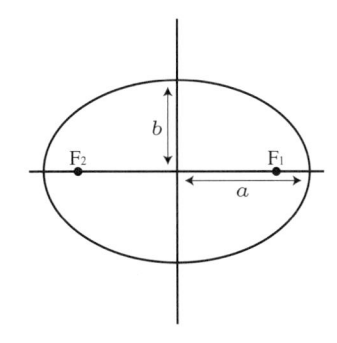

楕円　　　　　　　　　　　　　　　`ellipse,または,oval`

たまご形の平面図形で，対称軸 (axes of symmetry) である2つの直径をもっています．最長直径は長径 (the major axis) で，最短直径は短径 (the minor axis) です．

直径に沿って中心点 (the center) から楕円までの線分 (line segment) が長半径 (semimajor axis) で，短径に沿って中心から楕円までの線分は短半径 (semiminor axis) です．

さらに，楕円は2つの焦点 (two foci) F_1, F_2 をもっています．

● **面積** (the area of an ellipse (または，oval))

楕円の面積は次の公式で表されます．

$A = \pi ab$

ここで，a：長半径の長さ，b：短半径の長さ

読み方: The area of an ellipse equals π times a times b, where a is the length of the semimajor axis and b is the length of the semiminor axis.

注1） 円，楕円などが描く曲線を2次曲線 (quadratic curves) と呼びます．

ほかに2次曲線としては次があります．
- 双曲線：hyperbola
- 放物線：parabola

注2） 2次曲線には次の用語もしばしば使用されます．

- 接線：tangent (または，tangential line)
- 法線：normal (または，normal line)
- 漸近線：asymptote (または，asymptotic line)
- 離心率：eccentricity

注3） 公式 (formula) とは，数学上の恒等式 (identity) で，一般的な公式 (general rule) または法則 (law) のことです．

上記に記載した三角形，矩形，円，そのほかの面積や周囲の長さを表す式は，equation ではなく formula が使用されます．

8 立体図形
solid figures

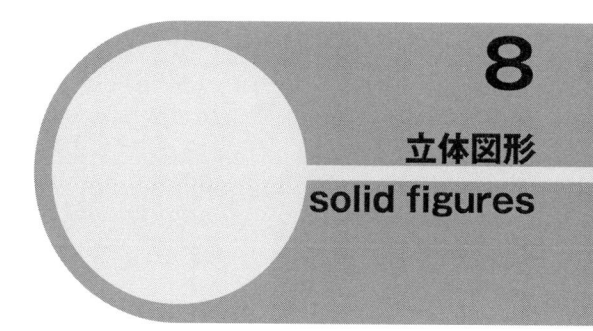

最も一般的な立体図形としては，下記があります．

球　　　　　　　　　　　　　　　　　　　　　　　sphere

点の軌跡 (locus) である閉じた表面で，すべての点は与えられた点 (中心点：the center) から固定距離 (半径) に存在します．

● **表面積** (the surface area of a sphere)

球の表面積 S は次の公式で表されます．

$S = 4\pi r^2$

ここで r は球の半径．

読み方：The surface area of a sphere equals 4 times π(pi) times r squared, where r is the radius of that sphere.

● **体積** (the volume of a sphere)

球の体積 V は次の公式で表されます．

$$V = \frac{4\pi r^3}{3}$$

ここで r は球の半径．

読み方：The volume of a sphere equals 4 times π(pi) times r

cubed divided by 3, where r is the radius of that sphere.

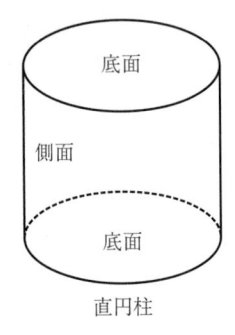

円柱   `cylinder`

底面
側面
底面
直円柱

円柱は，湾曲した側面 (curved lateral surface) をもつ2つの等しい平面，すなわち，底面 (two bases) より構成されます．

底面が側面と垂直な場合を直円柱 (right cylinder)，そうでない場合を斜円柱 (oblique cylinder) と呼びます．

円柱のもつ，2つの底面間の垂直距離 (perpendicular distance) を高さ (altitude) h とし，1つの底面の面積を A とすると，円柱の体積 V は次の公式で表されます．

$V = Ah$

読み方：The volume of a cylinder equals the area of the base times the altitude, where A is the area of a base and h is the altitude.

すい（錐） cone

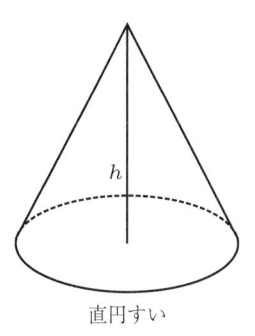

直円すい

　底面 (base) のすべての点と頂点 (vertex) を結ぶ，すべての線により形成される立体図形です．底面の形としては円形 (circle) や楕円形 (ellipse) などがあり，前者を円すい (circular cone)，後者を楕円すい (elliptical cone) と呼びます．閉曲線は円すいの準線 (directrix) です．底面に対して垂直な軸 (the axis perpendicular to the base) をもつ円すいを直円すい (right circular cone)，そのほかを斜円すい (oblique circular cone) と呼びます．すいの体積 (the volume of a cone) V は次の公式で表されます．

$$V = \frac{1}{3}hA$$

　ここで，h はすいの高さ (底面の平面から頂点への垂直距離) で，A は底面の面積です．

　読み方：The volume of a cone equals one-third times h times A, where h is the altitude of the cone which has the perpendicular distance from the plane of the base to the vertex and A is the area of the base.

多面体　　　　　　　　　polyhedron, 複数形：polyhedra

　平らな多面 (plane polygonal faces) よりなる平面をもつ，多面体のことです．多面体の中で，正多面体 (regular polyhedron) は等しい［または合同 (congruent)］な規則正しい多角形をもち，しかもそれで面を形成し，合同な多面角 (polyhedral angles) をもっています．

　凸正多面体 (convex regular polyhedra) には次の 5 種類があります．
- 四面体 (tetrahedron)：4 つの 3 角面
- 立方体 (cube)：6 つの 4 角面
- 八面体 (octahedron)：8 つの 3 角面
- 十二面体 (dodecahedron)：12 の 5 角面
- 二十面体 (icosahedron)：20 の 3 角面

● 四面体 (tetrahedron，複数形 tetrahedra)
　4 個の 3 角面 (triangular faces) をもつ立体図形です．

● 立方体 (cube)
　6 個の等しい 4 角面 (square faces) をもつ立体図形で，すべての面角 (face angles) が直角を形成します．

立方体の体積は次の公式で表されます．

$$V = a^3$$

ここで a は 1 つの稜（りょう［the length of an edge］）の長さです．

　読み方：The volume of a cube is a cubed, where a is the

length of an edge.

立方体は
　幅＝奥行＝高さ＝a
です．幅，奥行および高さがそれぞれ異なる長さの多面体を，長方形立方体 (rectangular solid) と呼びます．

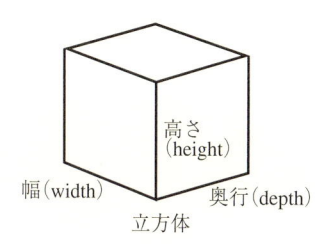

四面体

立方体　幅 (width)　奥行 (depth)　高さ (height)

角柱　　　　　　　　　　　　　　　　　　`prism`

角柱には，側面 (lateral edges) が底面 (bases) に対して直角をなす直角柱 (right prism) と，それ以外の斜角柱 (oblique prism) があります．
また角柱には次の種類があります．

● 三角柱 (triangular prism)
三角柱は2つの三角底辺 (triangular bases) と3つの側面 (lateral faces) をもっています．

● 四角柱 (quadrangular prism)
四辺形 (quadrilaterals) の底面をもっています．

● 五角柱 (pentagonal prism) は五角形の底面，および六角

柱 (hexagonal prism) は六角形の底面をもっています．

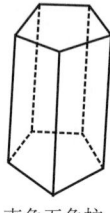

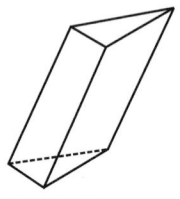

直角五角柱 　　斜角三角柱

角すい　　　　　　　　　　　　　　　　　　　`pyramid`

多角形 (polygon) の底面 (base) と，それとは同一面 (the same plane) にはない1つの点が共通頂点 (common vertex) をもつ，三角形の側面 (lateral faces) により構成される立体図形です．

底面に対して垂直な軸 (axis) をもつ角すいが直角すい (right pyramid) で，それ以外は傾斜角すい (oblique pyramid) です．

角すいの高さ (altitude) h は，底面から頂点までの垂直距離です．

● 体積 (the volume of a pyramid)

角すいの体積 V は次の公式で表されます．

$$V = \frac{1}{3}Ah$$

ここで A は底面積，h は高さです．

読み方：The volume of a pyramid is one-third times A times h, where A is the area of the base and h is the altitude.

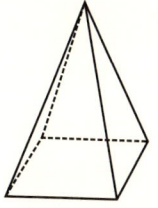

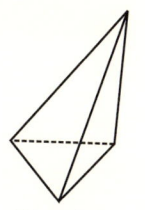

直角四角すい　　斜角三角すい

8　立体図形
solid figures

9 直交座標系
rectangular coordinate system

直交座標系

両軸 (axes) が直角に交わる座標系のことです．

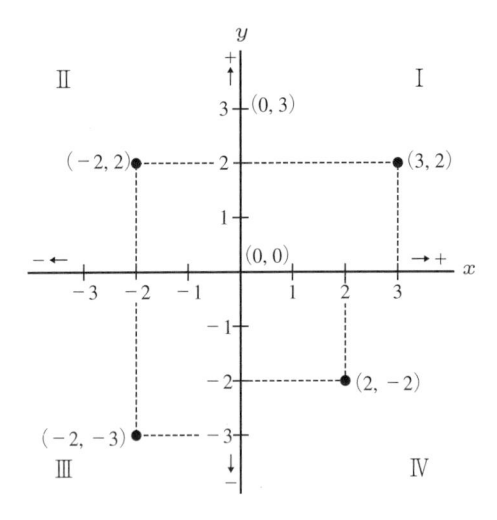

```
          y
     II   |    I
        3-(0, 3)
(-2, 2)●---- 2             ●(3, 2)
          1-
   ――――――(0,0)――――――→ + x
  -3 -2 -1 | 1  2  3
         -1-
         -2------●(2, -2)
(-2, -3)●----- -3
     III       IV
```

上の座標図 (coordinate graph) が示すように, 数直線 (number lines) 上の各点には

  …, -3, -2, -1, 0, 1, 2, 3, …

のような数が割り振られ (Each point on a number line is assigned a number), しかも平面 (plane) 上の各点には1対の数が割り振られています (Each point in a plane is assigned a pair of numbers.). これらの数は2本の交差線 (two intersecting lines) に対する, 点の配置を表しています.

上の座標図では, 直角に交わる2本の数直線 (two perpendicular number lines) が使用されています. これらを座標軸 (coordinate axes) と呼びます. そのうちの1つ, 水平な座標軸 (one axis) を x 軸 (x axis) と呼び, もう一方の垂直な座標軸を y 軸 (y axis) と呼びます. これら2本の数直線の交点 (the point of intersection) を, (座標の) 原点 (the origin) と呼び, 座標 (0, 0) [the coordinates (0, 0)] で表します.

x 軸で0より右側の数 (numbers to the right of 0) は正で, 左側の数は負となります. さらに y 軸で0より上の数 (numbers above 0) は正で, 下側の数 (numbers below 0) は負となります.

例えば, (3, 2) という1対の数のうち, 最初の数 (the first number) 3は x 座標 (x-coordinate) または横座標 (abscissa) に対する数, さらに2番目の数 (the second number) 2は y 座標 (y-coordinate) または縦座標 (ordinate) に対する数で, 座標 (3, 2) [the coordinates (3, 2)]を構成しています.

(3,2) というかっこ内の1対の数を, 順序づけされた1対の数 (ordered pair) と呼びます. 通常, 順序はアルファベット x, y の順に並んでいます.

座標図はさらに4つに区分され, これらを象限 (quadrants) と呼びます.

上図でⅠ, Ⅱ, Ⅲ, Ⅳの部分をそれぞれ第1象限 (the first

quadrant, または, quadrant Ⅰ), 第2象限 (the second quadrant, または, quadrant Ⅱ), 第3象限 (the third quadrant, または, quadrant Ⅲ), 第4象限 (the fourth quadrant, または, quadrant Ⅳ) と呼びます.

例えば, 第1象限では x, y ともに, 常に正です.

● そのほかの座標系

直交座標系以外にも次の座標系が存在します.

- ・カルテシアン座標系: Cartesian coordinate system
- ・極座標系: polar coordinate system
- ・慣性座標系: inertial coordinate system

構文例および読み方

⭕ 第1象限では, x は常に正で, y も常に正となります.

➡ In the first quadrant, x is always positive and y is always positive.

⭕ 点Aと点Bを下記の座標軸上で識別しなさい.

点A および B の座標はそれぞれ, (3, 2), (−2, −3) です.

➡ Identify the points of A and B on the coordinate graph below.

The coordinates at the points of A and B are 3 and 2 (3, 2) and minus 2 and minus 3 (−2, −3), respectively.

---

9 直交座標系
rectangular coordinate system

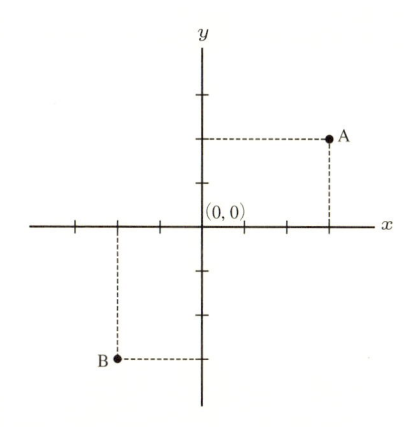

◐ 座標 (3, 14) は，1次方程式 $y = 5x - 1$ の解かどうか決定しなさい．

座標 (3, 14) はアルファベット順に並んでいますので，$x = 3$, $y = 14$ です．x に 3 を，y に 14 を代入します．
$$14 = 15 - 1$$
$$14 = 14$$
上式は正しい記述ですから，座標 (3, 14) は $y = 5x - 1$ の解です．

➡ Determine whether the ordered pair (または，point) (3,14) [3 and 14 (in the parentheses)] is a solution of the linear equation
$y = 5x - 1$
読み方：y equals 5 times x minus 1

The ordered pair is in alphabetical order, so $x = 3$ and $y = 14$. Substitute 3 for x and 14 for y.
$$14 = 15 - 1$$
$$14 = 14$$

This is a true statement. Therefore, the ordered pair (3,14) is a solution to $y=5x-1$.

この読み方が次の構文にも適用されます.

◐ 方程式 $y=-1$ に対して座標 ($\Box$, -1) を完成しなさい.
上式から y は -1 でなければなりませんので,未知の x 座標は 0 です.その結果 $(0,-1)$ となります.

➡ Complete the given ordered pair for the equation $y=-1$.
　　　($\Box$, -1)

Since y must equal -1, the missing x-coordinate is 0. Thus, the ordered pair $(0,-1)$ is the required solution.

◐ 3つの座標 $(0,0)$, $(2,4)$, および $(2,0)$ をプロットしなさい.

➡ Plot the three coordinates (または,ordered pairs, points) $(0,0)$, $(2,4)$ and $(2,0)$.

◐ 方程式 $y=2x$ をグラフに描きなさい.

$x=1$ のとき $y=2$, $x=2$ のとき $y=4$, $x=3$ のとき $y=6$ となります.それゆえ,$y=2x$ の解として次の座標が得られます.
これらの点をプロットして直線で結びます.

➡ Graph the equation $y=2x$.

If $x=1$, $y=2$. If $x=2$, $y=4$. If $x=3$, $y=6$.
The following ordered pairs are obtained as solutions of $y=2x$.

   (1, 2) (2, 4) (3, 6)

Plot these points and connect them with a straight line.

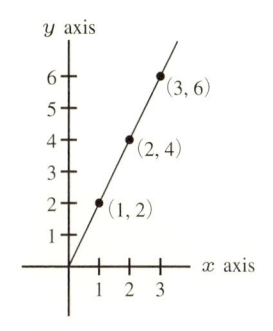

◐ $x=3$ をグラフに描くには,まず点 (3, 0) の位置を求め,次に (3, 0) を通る垂直線を引きます.

➡ To graph $x=3$, locate the point (3, 0). Then, draw a vertical line through (3, 0).

◐ (1, 2) および (3, 5) を通る直線の傾きを求めなさい.

点 (1, 2) および (3, 5) を含む直線の傾きは,次のように求められます.

$$\text{傾き} = \frac{5-2}{3-1} = \frac{3}{2}$$

この場合,この直線は正の傾きです.
傾きが正 (または負) の場合,左から右に向かって上昇 (または下降) します.

➡ Find the slope of a straight line through the points (1, 2) and (3,

9 直交座標系
rectangular coordinate system

5).

The slope of the line containing the points (1, 2) and (3, 5) is as follows.
$$\text{Slope} = \frac{5-2}{3-1} = \frac{3}{2}$$
This line has a positive slope.
When the slope of a line is positive (または, negative), it rises (または, falls) from left to right.

式　　　　　　　　　　　　　　　　　　　　`expression`

多項式，方程式などにおける，記号表示の数式 (any mathematical form expressed symbolically, as in a polynomial, equation, etc.) です．多項式 (a polynomial)，方程式 (または，等式) (an equation)，不等式 (an inequality)などの中に存在します．

例えば，方程式
$$x^2 - 3x - \frac{5}{2} = 0$$
の場合，
$$x^2 - 3x - \frac{5}{2}$$
がこの方程式を構成する式 (expression) です．x^2, $-3x$, $-5/2$ はそれぞれ項 (a term) と呼ばれます．

この場合，最初の項 (the first term) は x^2, 最後の項 (the last term) は $-5/2$ です．また項数 (the number of terms) は x^2, $-3x$ および $-5/2$ の3個です．

同様に，$2x+6$, $x-4$ などもそれぞれ式 (expression) です．

式が集まって多項式，方程式または不等式などが構成されます．

式に使用される代表的用語は下記のとおりです．

- 定数 (constant)：一般には数．（例） 5, 8
- 変数 (variable)：未知数を表すために使用される文字．（例） x, y, z, a, b, c
- 項 (term)：数や数と変数の積，数とべき乗された変数の積など．（例） $4, 2xy, 5x^3$
- (数値) 係数 [(numerical) coefficient]：項中の数の部分．（例） $2xy$ 中の 2
- 同類項 (similar or like terms)：同一変数および指数をもつ項．（例） $2x^2$ と $3x^2$，$4x$ と $5x$，x^2y と $5x^2y$
- 項の次数 (degree of a term)：その項中にある変数の指数の合計．（例） $6x^2y^3$ の項の次数は 5，$2x(=2x^1)$ の項の次数は 1
- 各項の最高次数 (変数が 1 つの場合) (highest degree of any term)：（例） $4x^3-2x$ の場合は 3．x^5-3x^3 の場合は 5．

構文例および読み方

⭕ 式を簡略化するために，その式の中の項数を通分します．

➡ To simplify expressions, we reduce the number of terms in the expression.

⭕ できるだけ完全に下記式を因数分解して下さい．
$$4x^3+10x^2y-24xy^2$$

➡ Factor the following expression as completely as possible.
   $4x^3 + 10x^2y - 24xy^2$
   読み方: 4 times x cubed plus 10 times x squared times y minus 24 times x times y squared

◯ $\sqrt{25-9}$ は，式 $25-9$ の平方根で $\sqrt{16}=4$ に等しい．

➡ $\sqrt{25-9}$ is the square root of the expression $25-9$ and is equal to $\sqrt{16}=4$.

◯ 式 $3x$, $(a+b)x$ および $2xyz$ 中の x の係数はそれぞれ，3, $(a+b)$ および $2yz$ です．係数は通常は定数です．

➡ The coefficients of x in the expressions $3x$, $(a+b)x$, and $2xyz$ are respectively 3, $(a+b)$, and $2yz$. A coefficient is usually a constant.

多項式　　　　　　　　　　　　　　　polynomial

多項式は，項の数から単項式 (monomial), 2 項式 (binomial), 3 項式 (trinomial) などに分類され，さらに次数から線形 (linear), 2 次 (quadratic), 3 次 (cubic), 4 次 (biquadratic), 5 次 (quintic) 式などに分類されます．

● 単項式 (monomial)

　1 つの項 (term) のみを含む式 (expression) のことです．
(例)　$9x$, $4a^2$, $4x^2$

- **2 項式 (binomial)**

  2つの項 (two terms) を含む式 (expression) のことです．
  (例)　$x+y,\ 4x^3-2x$

- **3 項式 (trinomial)**

  3つの項 (three terms) を含む式 (expression) のことです．
  (例)　$y^2+9y+8,\ x^5-2x^3-6x$

構文例および読み方

⊃ 次の式は 2 項式か，3 項式か識別しなさい．
　　$7x^2-5xy+3y^4$

上式は 3 項式です．

➡ Identify whether the following expression is binomial or trinomial.
　　$7x^2-5xy+3y^4$

The above expression is a trinomial.

⊃ 次の 2 つの 3 項式を加算しなさい．
　　$3x^2+4x+7,\ 5x^2+2x+12$

　　$(3x^2+4x+7)+(5x^2+2x+12)=8x^2+6x+19$

➡ Add the following two trinomials.
　　$3x^2+4x+7,\ 5x^2+2x+12$
　　$(3x^2+4x+7)+(5x^2+2x+12)=8x^2+6x+19$
　　読み方: 3 times x squared plus 4 times x plus 7, these

three terms in the parentheses plus 1 on the outside of the parentheses times 5 times x squared plus 2 times x plus 12 on the inside equals 8 times x squared plus 6 times x plus 19.

◐ 3項式 $16x^4 - 8x^3 - 2x$ を単項式 $2x$ で除算しなさい.

$$\frac{16x^4 - 8x^3 - 2x}{2x} = 8x^3 - 4x^2 - 1$$

➡ Divide the trinomial $16x^4 - 8x^3 - 2x$ by the monomial $2x$.

$$\frac{16x^4 - 8x^3 - 2x}{2x} = 8x^3 - 4x^2 - 1$$

方程式, 等式   `equation`

2つの数式が等しいという表現です (a statement that two mathematical expressions are equal).

(例)

$2x + 6 = x - 4$ 線形方程式 (linear equation)

$x^2 - 3x - \frac{5}{2} = 0$ 非線形方程式 (nonlinear equation)

等号 [equality symbol (または, equal sign)]

$\boxed{2x+6}$ = $\boxed{x-4}$

left side   right side
(左辺)   (右辺)

すなわち, 上例では方程式の左辺の式 (expression) $2x+6$

と，右辺の式 (expression) $x-4$ が等号で結ばれています．

方程式 (equation) には，

- 1次方程式 (linear equation) (例) $ax=b$
- 2次方程式 (quadratic equation) (例) $ax^2+bx+c=0$
- 3次方程式 (cubic equation) (例) $y^3+py^2+qy+r=0$
- 4次方程式 (biquadratic equation)

$$(例)\ y^4+py^3+qy^2+ry+s=0$$

などがあります．

構文例および読み方

◯ 例えば，$x+3y+2z=7$ は 3 変数の線形方程式です．

➡ For example, $x+3y+2z=7$ is a linear equation in three variables.

◯ 次の方程式を解きなさい．
   $7y=6y-3$

➡ Solve the following equation.
   $7y=6y-3$

◯ かっこを外し，さらに同類項を加算して，方程式の左辺を簡略化しなさい．

➡ Simplify the left side of the equation by removing parentheses and adding similar terms.

不等式　　　　　　　　　　　　　　　　　　inequality

　1つの辺の式が，ほかの辺の式より値の上で大きい (または，等しい)，もしくは小さい (または，等しい) を表す記述 (a statement) のことです．

(例)
$$4(x+1)-7+2x > 5+2x,\ 4(x+1)-7+2x \geq 5+2x$$
$$x-1 < 2x^2+1,\ x-1 \leq 2x^2+1$$

構文例および読み方

○ 次の不等式を解きなさい．
$$3x < 12$$

➡ Solve the following inequality.
$$3x < 12$$

○ われわれは，線形方程式を解くのとほぼ同じ方法で，線形不等式を解きます．

➡ We solve linear inequalities almost the same way we solve linear equations.

10 微分および積分
derivatives and integrals

微分　derivative,または,differential,differentiation

独立変数 x (independent variable) に対する, 関数 (function) $y=f(x)$ の変化率 (the rate of change) のことです.

関数 $y=f(x)$ の場合, その微分 (derivative) は次のようになります.

(1次) 微分　　　　　　　　　　　　　　derivative

独立変数 x に対する関数 $y=f(x)$ の微分, すなわち

$\dfrac{dy}{dx}$　（または, y', $\dfrac{df(x)}{dx}$）

読み方: The derivative of the function $y=f(x)$ (y being equal to f of x) with respect to x.

注) 1次微分とはいわず, 単に微分といいます.

2次微分　　　　　　　　　　　second-order derivative

独立変数 x に対する関数 $y=f(x)$ の2次微分, すなわち

$$\frac{d^2y}{dx^2} \quad (\text{または}, \ y'', \ \frac{d^2f(x)}{dx^2})$$

読み方: The second-order derivative of the function $y=f(x)$ with respect to x.

n 次微分　　　　　　　　　　　　　　　　nth-order derivative

独立変数 x に対する関数 $y=f(x)$ の n 次微分, すなわち

$$\frac{d^n y}{dx^n} \quad (\text{または}, \ \frac{d^n f(x)}{dx^n})$$

読み方: The n th-order derivative of the function $y=f(x)$ with respect to x.

構文例および読み方

◯ 次の関数を微分しなさい.

$$y=x^4 \quad \rightarrow \quad \frac{dy}{dx}=4x^3$$

$$y=x^3+3x^2+2x+1 \quad \rightarrow \quad \frac{dy}{dx}=3x^2+6x+2$$

➡ Take the derivative of the following functions.
　　$y=x^4$
　　読み方: y equals x to the fourth power.

The derivative of x^4 with respect to x equals $4x^3$.
　　読み方: 4times x cubed.

$y=x^3+3x^2+2x+1$
読み方: y equals x cubed plus 3 times x squared plus 2 times x plus1.

The derivative of x^3+3x^2+2x+1 with respect to x equals $3x^2+6x+2$.

読み方：3 times x squared plus 6 times x plus 2.

◯ 次の関数を 2 次微分しなさい．
$$y = x^4 \rightarrow \left(\frac{dy}{dx} = 4x^3\right) \rightarrow \frac{d^2y}{dx^2} = 12x^2$$

➡ Take the second-order derivative of the following function.
$y = x^4$
読み方：y equals x to the fourth power.

The derivative of x^4 with respect to x equals $4x^3$.
読み方：4 times x cubed.

The second-order derivative of x^4 with respect to x equals $12x^2$.
読み方：12 times x squared.

偏微分　　　　　　　　　　　　　　partial derivative

　1 つ以上の変数 (variables) を含む関数 (function) において，それらの変数の 1 つに着目したときの，その関数の変化率 (the rate of change) のことです．そのほかの変数は定数として取り扱われます．
　例えば，関数 $u = f(x, y, z, \cdots)$ の場合，変数 x に対する関数 u の偏微分とは，x のみが変化した場合の，関数 u の変化率のことです．
　この場合，y および z は変数ではなく，定数として取り扱われます．

(1次) 偏微分   `partial derivative`

独立変数 x に対する関数 $u=f(x, y)$ の偏微分,すなわち

$$\frac{\partial u}{\partial x} \quad (または, \ \frac{\partial f(x, y)}{\partial x})$$

読み方: The partial derivative of the function $u=f(x, y)$ (u being equal to f of x and y) with respect to x.

注) 1次偏微分とはいわずに,単に偏微分といいます.

n 次偏微分   `nth-order partial derivative`

独立変数 x に対する関数 $u=f(x, y)$ の n 次偏微分,すなわち

$$\frac{\partial^n u}{\partial x^n} \quad (または, \ \frac{\partial^n f(x, y)}{\partial x^n})$$

読み方: The nth-order partial derivative of the function $u=f(x, y)$ with respect to x.

注) 上記は,$u=f(x, y)$ の x を独立変数とし,y を定数とした場合の偏微分について説明しましたが,x と y をともに独立変数として偏微分する場合もあります.この場合,記号で

$$\frac{\partial^2 u}{\partial x \partial y}$$

と書き,これは,まず x に関し,次に y に関して微分して得られた関数 $u=f(x, y)$ の偏微分のことで,次の式で表されます.

$$\frac{\partial^2 f(x, y)}{\partial x \partial y}$$

読み方: The second-order partial derivative of the function $u = f(x, y)$ (u being equal to f of x and y) obtained by differentiating first with respect to x and then with respect to y.

構文例および読み方

◯ $V = \pi r^2 h$ を r に対して偏微分しなさい．

$$\frac{\partial V}{\partial r} = 2\pi r h$$

➡ Take the partial derivative of $V = \pi r^2 h$ with respect to r.

読み方: The partial derivative of $\pi r^2 h$ with respect to r equals 2 times π times r times h.

◯ $V = \pi r^2 h$ を h に対して偏微分しなさい．

$$\frac{\partial V}{\partial h} = \pi r^2$$

➡ Take the partial derivative of $V = \pi r^2 h$ with respect to h.

読み方: The partial derivative of $\pi r^2 h$ with respect to h equals π times r squared.

◯ $\dfrac{\partial^2 V}{\partial x^2} + \dfrac{\partial^2 V}{\partial y^2} + \dfrac{\partial^2 V}{\partial z^2} = 0$

➡ The second-order partial derivative of V with respect to x plus the second-order partial derivative of V with respect to y plus the second-order partial derivative of V with respect to z equals zero.

➲ $\dfrac{1}{r^2}\dfrac{\partial}{\partial r}\left(r^2\dfrac{\partial V}{\partial r}\right)$

➡ One divided by r squared times the partial derivative with respect to r of r squared times the partial derivative of V with respect to r.

積分　　　　　　　　　　　`integral,`または`,integration`

$F(x)$ を x の関数 (function) として, $F(x)$ の微分を $f(x)$ とするとき, $F(x)$ を $f(x)$ の積分と呼び, 次の式で表します. 特定の積分区間のない積分を不定積分 (indefinite integral) と呼びます.

$$F(x)=\int f(x)\,dx$$

読み方: Capital F of x equals the (indefinite) integral of f of x with respect to x.

独立変数 x の 2 つの値に対する, 2 つの積分の差を定積分 (definite integral) と呼び, 次の式で表します.

$$F(b)-F(a)=\int_a^b f(x)\,dx$$

読み方: Capital F of b minus capital F of a equals the integral taken between the values b and a (または, from the value a to the value b) of f of x with respect to x.

注) 1 重積分 (single integral) とはいわずに, 単に積分とい

います.

多重積分   `multiple integral`

2回,またはそれ以上連続して積分を行うことを多重積分と呼び,まず1つの変数を積分し,ほかは定数として扱い,次に別の変数が積分され,ほかを定数として取り扱います.

2重積分 (double integral) は下記のように書きます.

$$\iint f(x, y)\, dxdy = \int \left[\int f(x, y)\, dx \right] dy$$

読み方: The double (indefinite) integral of f of x and y obtained by integrating first with respect to x and then with respect to y.

3重積分は,triple integral と呼びます.

構文例および読み方

● $\int x^2\, dx$ を求めなさい.

$$\int x^2\, dx = \frac{x^{2+1}}{2+1} + C \quad (n \neq -1,\ \text{および},\ C = \text{定数})$$
$$= \frac{x^3}{3} + C$$

➡ Take the (indefinite) integral of x^2 (x squared) with respect to x.

読み方: The (indefinite) integral of x^2 with respect to x equals x cubed divided by 3, the divisor plus C, where n is not equal to minus 1 and C is a constant.

● $\int_2^4 x\,dx$ を求めなさい.

$$\int_2^4 x\,dx = \left(\frac{4^2}{2}+C\right) - \left(\frac{2^2}{2}+C\right)$$
$$= 8+C-2-C$$
$$= 6$$

➡ Take the (definite) integral taken between the values 4 and 2 of x with respect to x.

> 読み方: The integral taken between the values 4 and 2 of x with respect to x equals 8 plus C minus 2 minus C, namely, 6.

11 三角関数
trigonometric functions

三角関数　　　　　　　　　　　　trigonometric function

∠c が直角の (∠c as the right angle) 直角三角形 (right-angled triangle) ABC において，角 A, B, C と対向した辺 (side) の長さが a, b, c とすると，三角関数は下記のとおりです．

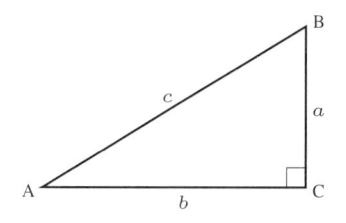

- 正接 (tangent)： $\tan A = \dfrac{a}{b}$
- 正弦 (sine)： $\sin A = \dfrac{a}{c}$
- 余弦 (cosine)： $\cos A = \dfrac{b}{c}$
- 余接 (cotangent)： $\cot A = \dfrac{b}{a}$ (ctnA とも書きます)
- 余割(ヨカツ) (cosecant)： $\csc A = \dfrac{c}{a}$ (cosecA とも書きます)

- 正割 (secant)： $\sec A = \dfrac{c}{b}$

上の cot A, csc A および sec A はそれぞれ, tan A, sin A および cos A の逆数 (reciprocal) となります．

$$\cot A = \dfrac{1}{\tan A}$$

$$\csc A = \dfrac{1}{\sin A}$$

$$\sec A = \dfrac{1}{\cos A}$$

構文例および読み方

◯ $\sin A = \dfrac{a}{c}$

→ The sine of A equals a divided by c.

◯ $\csc A = \dfrac{c}{a}$

→ The cosecant of A equals c divided by a.

◯ $\cot A = \dfrac{1}{\tan A}$

→ The cotangent of A equals the reciprocal of the tangent of A.

◯ $\cos(-\alpha) = +\cos\alpha$

→ The cosine of minus α equals plus the cosine of α.

◯ $\sin\alpha \cos\beta = \dfrac{1}{2}\{\sin(\alpha+\beta) + \sin(\alpha-\beta)\}$

→ The sine of α times the cosine of β equals one-half on the outside of the braces times the sine of α plus β, both angles in the parentheses plus the sine of α minus β, both angles in the parentheses on the inside of the braces.

◐ $\sin \frac{\alpha}{2} = \pm \sqrt{\left(1 - \cos \frac{\alpha}{2}\right)}$

➡ The sine of α over 2 equals plus or minus the square root of 1 minus the cosine of α over 2, both radicands in the parentheses.

◐ $\sin^2 \alpha + \cos^2 \alpha = 1$

➡ The sine squared of α plus the cosine squared of α equals 1.

逆三角関数

     inverse trigonometric function, または,
       antitrigonometric function

三角関数の逆元 (the inverse of trigonometric functions) のことです.

例えば $y = \tan x$ の場合, 逆三角関数は $x = \tan^{-1} y$ と書きます.

- アークサイン (arcsine): $\sin^{-1} y$
- アークコサイン (arccosine): $\cos^{-1} y$
- アークタンジェント (arctangent): $\tan^{-1} y$

逆三角関数は, 下記の制限された範囲内に存在する主値 (principal values) をもつ, 1価関数 (single-valued function) とみなされます.

x の範囲

$$x = \sin^{-1} y \quad : -\frac{\pi}{2} \leq x \leq \frac{\pi}{2}$$

$$x = \cos^{-1} y \quad : 0 \leq x \leq \pi$$

$$x = \tan^{-1} y \quad : -\frac{\pi}{2} \leq x \leq +\frac{\pi}{2}$$

構文例および読み方

● $\sin^{-1} a + \cos^{-1} a = \dfrac{\pi}{2}$

➡ The arcsine of a plus the arccosine of a equals π over 2.

● $\sin^{-1} a = \pm \cos^{-1} \sqrt{(1-a^2)}$

➡ The arcsine of a equals plus or minus the arccosine of the square root of 1 minus a squared, both radicands in the parentheses.

双曲線関数　　　　　　　　　　hyperbolic function

代表的な双曲線関数としては次のものがあります．

- 双曲線正弦 (hyperbolic sine)：$\sinh x$
- 双曲線余弦 (hyperbolic cosine)：$\cosh x$
- 双曲線正接 (hyperbolic tangent)：$\tanh x$

構文例および読み方

◯ $\sinh x = \dfrac{1}{2}(e^x - e^{-x})$

➡ The hyperbolic sine of x equals one-half on the outside of the parentheses times e to the xth power minus e to the minus xth power on the inside.

◯ $\sinh(-x) = -\sinh x$

➡ The hyperbolic sine of minus x equals minus the hyperbolic sine of x.

逆双曲線関数   inverse hyperbolic function

双曲線関数の逆元 (the inverse of hyperbolic function) で, 代表的な逆双曲線関数としては下記があります.

・逆双曲線正弦 (hyperbolic arcsine): $\sinh^{-1} x$
・逆双曲線余弦 (hyperbolic arccosine): $\cosh^{-1} x$
・逆双曲線正接 (hyperbolic arctangent): $\tanh^{-1} x$

構文例および読み方

◯ $\sinh^{-1} x = \ln\left[x + \sqrt{(x^2 - 1)}\right]$

➡ The hyperbolic arcsine of x equals the logarithm of x plus the square root of x squared minus 1, both radicands in the parentheses to the base e.

◐ $\frac{1}{2} a^2 \cosh^{-1}\frac{x}{a}$

➡ One-half times a squared times the hyperbolic arccosine of x over a.

12 級数
progressions

　一連の数 (a sequence of numbers) のことで，これらの2つの連続した項の間 (between two consecutive terms) には，一定の関係 (a constant relation) が存在します．一般的級数には次の級数があります．

等差数列または算術数列
`arithmetic progression,`または，`arithmetic sequence`

　最初の項を除く各項が，一定の量 (公差：common difference) だけ，その前の項と異なる数列 (sequence) のことです．いま，最初の項を a，公差を d とすると，級数は次の形式となります．

$$a,\ a+d,\ a+2d,\ a+3d,\ \cdots,\ a+(n-1)d$$
　↓　　　　　　　　　　　　　↓
最初の項　　　　　　　　第 n 次の項
(the first term)　　　　　(the nth term)

　最初の n 項の合計は下記式となります．

$$na+\frac{1}{2}n(n-1)d$$

構文例および読み方

○ 最初の n 項の合計は次の式となります.

$$na+\frac{1}{2}n(n-1)d$$

➡ The sum of the first n terms is given by the following expression.

$$na+\frac{1}{2}n(n-1)d$$

読み方: n times a plus one-half times n on the outside of the parentheses times n minus 1 on the inside times d.

○ $1+2+3+\cdots+n=\dfrac{n(1+n)}{2}$

➡ 1 plus 2 plus 3 plus, and so on, plus n equals n on the outside of the parentheses times 1 plus n on the inside divided by 2.

等比級数

`geometric progression,` または, `geometric sequence`

最初の項を除く,各項と次の項の比 [the ratio of each term (except the first term) to the preceding term] が,一定 (公比: common ratio) な数列のことです.

最初の項を a, 公比を r として,級数は次の形式となります.

$$a,\ ar,\ ar^2,\ ar^3,\ \cdots,\ ar^{n-1}$$

↓   ↓

最初の項   第 n 次の項

(the first term)   (the nth term)

$r \neq 1$ として,最初の n 項の合計は,

$$\frac{a(1-r^n)}{(1-r)}$$

構文例および読み方

◯ $r \neq 1$ として,最初の n 項の合計は次の式で表されます.

$$\frac{a(1-r^n)}{(1-r)}$$

➡ If $r \neq 1$ (r is not equal to 1), the sum of the first n terms is given by the following expression.

$$\frac{a(1-r^n)}{(1-r)}$$

読み方:a on the outside of the parentheses times 1 minus r to the nth power on the inside divided by 1 minus r, both terms in the parentheses.

◯ $1^2 + 2^2 + 3^2 + \cdots + n^2 = \dfrac{n(1+n)(1+2n)}{6}$

➡ 1 squared plus 2 squared plus 3 squared plus, and so on, plus n squared equals n on the outside of the parentheses times 1 plus n on the inside times 1 plus $2n$, both terms in the parentheses divided by 6.

---

12 級数

progressions

注1) 数列 (sequence) とは，ある規則や法則にしたがって形成される連続した項 (a_1, a_2, a_3, a_4,…)のことで，例えば次が挙げられます．

　　1, 4, 9, 16, 25, …

　　1, -1, 1, -1, 1, …

　　$\dfrac{x}{1!}, \dfrac{x^2}{2!}, \dfrac{x^3}{3!}, \dots$

したがって，上記で説明した級数も広い意味では，これら数列の中に含まれます．

注2) 上記の級数に加え，調和級数 (harmonic progression, または，harmonic sequence) もあります．

13

順列および組合せ
permutation and combination

順列および組合せで使用される用語には，次のものがあります．

n の階乗

$n! = n(n-1)(n-2) \cdots 3 \cdot 2 \cdot 1$
$n!$
読み方：n factorial, または, factorial n

順列 permutation

選択の順序が重要な場合で，n 個の識別可能な目標物から $r \leq n$ 個の目標物を選択する場合，何通りの選択方法があるかを求めることです．

次の記号で表されます．

$_nP_r$（または，^nP_r）$= n(n-1)(n-2)\cdots(n-n+1)$

読み方：The number of permutations of n objects* taken r at a time equals n times n minus 1, both terms in the parentheses times n minus 2, both terms in the parentheses times, and so on, times n minus r plus 1, these three terms in the parentheses.

上式で $r=n$ の場合は，$_nP_n = n!$ となります．

* objects は，具体的な物が指定された場合，その物の名称で呼びます．構文例および読み方を参照．

組合せ combination

選択の順序を無視する場合で，n 個の識別可能な項目から r 個の異なった項目を選択する場合，何通りの組合せがあるかを求めることです．次の記号で表されます．

$$C(n, r) \text{ (または，} {}_nC_r, C_r^n) = \frac{n!}{r!(n-r)!}$$

読み方：The number of combinations of n things* taken r at a time equals factorial n divided by factorial r multiplied by factorial n minus r, both terms in the parentheses.

* things は，具体的な物が指定された場合，その物の名称で呼びます．(構文例および読み方を参照．)

構文例および読み方

○ 積 $4 \times 3 \times 2 \times 1$ を，4! と書きます．

➡ The product $4 \times 3 \times 2 \times 1$ (4 times 3 times 2 times 1) can be written 4! (4 factorial).

⊃ 3個の異なったコーヒーカップを棚に横1列に並べるには，何通りの異なった方法がありますか？

コーヒーカップの並べ順序はその前の選択方法に影響されるので，異なった並べ方は，$_3P_3 = 3!$，すなわち，$3 \times 2 \times 1 = 6$ 通りの方法となります．

→ How many different ways are there to arrange three different coffee cups in a row on a shelf ?

Since the order of the coffee cups is affected by the previous choices, the equation is as follows.
$_3P_3 = 3! = 3 \times 2 \times 1 = 6$
読み方: The number of permutations of 3 coffee cups taken 3 at a time equals 3 factorial, namely, 6.
This means there are six permutations.

⊃ a, b, c, d の4文字から同時に2文字を取り出す場合，何通りの組合せがありますか？

$n=4, r=2$ より，式は次のようになります．
$$C(4,2) = \frac{4!}{2!(4-2)!} = \frac{4!}{2!(2)!} = \frac{4 \cdot 3 \cdot 2 \cdot 1}{2 \cdot 1 (2 \cdot 1)} = 6$$

→ How many possible combinations of a, b, c and d taken 2 at a time are there ?

Since $n=4$ and $r=2$, the equation is as follows.
$$C(4,2) = \frac{4!}{2!(4-2)!} = 6$$
読み方: The number of combinations of four letters taken two at a time equals factorial 4 divided by factorial 2 mul-

> tiplied by factorial 4 minus 2, both terms in the parentheses, namely, 6.

注) $_3P_3$ や $C(4, 2)$ の読み方は，3 や 4 が何を指すかで変わります．上文では $_3P_3$ の 3 は 3 個のコーヒーカップ，$C(4, 2)$ の 4 は letters となり，$_3P_3$ や $C(4, 2)$ の 3, 4 が人なら three または four people, 野球選手なら three または four baseball players となります．そのつど，読み方が変わります．

14 和,合計または総和
summation or sum

和,合計または総和

一例として,収束級数 (convergent series) の合計 (sum) を求めます.

この場合の合計を表す記号としては,Σ(sigma) が使用されます.

いま,数列 (sequence) の第1の項 (the first term) a_1 から第 N 項まで a_n が合計されるものとすると,次の式で表されます.

$$\sum_{n=1}^{N} a_n$$

読み方: The sum of a (with the) sub(-script) n (on the right) from n being equal to 1 to N.

無限項数 (infinite number of terms) の合計は,次のように書きます.

注) a_{12} は簡単に a sub one two と読みます.

$$\sum_{n=1}^{\infty} a_n \quad (\sum a_n \text{ とも書きます.})$$

読み方: The sum of a sub n from n being equal to 1 to infinity (または, The sum of a sub n).

構文例および読み方

◌ $f(x) = \sum_{-\infty}^{\infty} a_n (z-z_0)^n$

➡ f of x equals the sum of a sub n on the outside of the parentheses times z minus z sub zero on the inside of the parentheses to the nth power from minus infinity to infinity.

◌ $\sum (y_i - a - bx_i)^2$

➡ The sum of y sub i minus a minus b times x sub i, these three terms in the parentheses to the second power.

◌ $I = \sum_{i=1}^{n} m_i r_i^2$

➡ I equals the sum of m sub i times r sub i to the second power from i being equal to 1 to n.

14 和, 合計または総和
summation or sum

15 行列式および行列
determinant and matrix

行列式　　　　　　　　　　　　　determinant

1組の数量，または要素 [a set of quantities (または, elements)] の合計，または積を正方形の配列 (square array) で表す1つの方法です．

(例)
$$\begin{vmatrix} a & b \\ c & d \end{vmatrix}$$

各要素の水平な並びが行 (rows) で，垂直な並びが列 (columns) です．

行 (または列) の数が行列式の次数 (the order of a determinant) で，上の例では2となります．

上部左側から (from top left), 下部右側への (bottom right) 対角線を主対角線 (the leading or principal diagonal) と呼びます．上例では a, d です．他方は2次対角線 (the secondary diagonal) と呼びます．上例では b, c です．

2次行列式 (second-order determinant) の値は $ad-bc$ です．

第 1 行 (the first row) → $\begin{vmatrix} a & b \\ c & d \end{vmatrix} = ad - bc$
第 2 行 (the second row) →

　　　　　　　　　　　　　　→ 第 2 列 (the second column)
　　　　　　　　　　　　　　→ 第 1 列 (the first column)

構文例および読み方

● $\begin{vmatrix} 1 & 2 & 3 \\ 4 & 1 & 2 \\ 6 & 5 & 4 \end{vmatrix}$

➡ The (third-order) determinant of the square matrix with 1, 2 and 3 in the first row, 4, 1 and 2 in the second row, and 6, 5 and 4 in the third row.

● $\begin{vmatrix} 1 & 2 \\ 5 & 4 \end{vmatrix} = 4 - 10 = -6$

➡ The (second-order) determinant of the square matrix with 1 and 2 in the first row, and 5 and 4 in the second row equals 4 minus 10, and equals minus 6.

行列　　　　　　　　　　　　　　　　　　`matrix`

　1 組の数量 (または要素) (a set of quantities or elements) を，矩形の配列 (rectangular array) で表す 1 つの方法です．

　配列は丸形中かっこ ( ) で閉じます．

　各要素の水平な並びが行 (rows) で，垂直な並びが列 (columns) です．

行列の次数は $m \times n$ で表されます．ここで m は行の数，n は列の数です．

$$\begin{pmatrix} 1 & 2 & 3 \\ 2 & 3 & 1 \end{pmatrix}$$
→ 第1行 (the first row)
→ 第2行 (the second row)
↳ 第3列 (the third column)
↳ 第2列 (the second column)
↳ 第1列 (the first column)

構文例および読み方

● $\begin{pmatrix} a & b \\ c & d \\ e & f \end{pmatrix}$

➡ The matrix with a and b in the first row, c and d in the second row and e and f in the third row.

16 そのほかの記号
other symbols and signs

そのほかの記号

いままで述べた以外に，しばしば使用される記号について，ここで記載します．

- $\sim$: 〜と相似, is similar to 〜
   〜の否定, the negation of 〜
- $\propto$: 〜に比例, is proportional to 〜
- $\therefore$: それゆえ, therefore
- Π : 積, product
- $\cup$: 2集合の和集合, union of two sets
- $\cap$: 2集合の共通集合, intersection of two sets
- $\subset$: 〜に含まれる, is included in 〜
   〜の部分集合, is a subset of 〜
- $\supset$: 部分集合として含む, contains as a subset
- $\in$: 〜の要素, is an element of 〜
- $\notin$: 〜の要素ではない, is not an element of 〜
- $|\ |$: 絶対値記号, natation of absolute value
- $\lim$: (関数，数列の) 極限, limit

使用例および読み方

- $A \in F$

→ A is an element of F.

- $A \subset B$

→ A is included in B.

- $\prod_{1}^{\infty} T_n$

→ The product of capital T sub n from 1 to infinity.

- $|-6|$

→ The absolute value of minus 6 is 6.

- $|2-6| > 3$

→ The absolute value of 2 minus 6 is greater than 3.

- $\lim_{x \to \infty} \left(\dfrac{1}{x} \right) = 0$

→ The limit of 1 over x as x tends to infinity is zero.

参考文献

LONGMAN DICTIONARY OF ENGLISH LANGUAGE AND CULTURE, Second Edition
Addison Wesley Longman Limited, second edition, 2000.

OXFORD WORDPOWER DICTIONARY for intermediate learners of English
edited by Sally Wehmeier, Ninth impression, Oxford University Press, 1995.

DICTIONARY OF MATHEMATICS
by John Daintith and R. D. Nelson, First edition, The Penguin, 1989.

MATH REVIEW FOR STANDARDIZED TESTS
by Jerry Bobrow, Ph.D., Wiley Publishing, Inc., 1985.

ELEMENTARY ALGEBRA by Joan Dykes, Ph.D., Harper Collins Publishers, 1991.

PROGRESSIVE
English-Japanese Dictionary, third edition, SHOGAKUKAN, 1999.

NEW JAPANESE-ENGLISH DICTIONARY
by Koh Masuda, editor in chief, fourth edition, KENKYUSHA, 1974.

NEW ENGLISH-JAPANESE DICTIONARY
by Yoshio Koine, editor in chief, fifth edition, KENKYUSHA, 1980).

著者略歴

鵜沼　仁（うぬま・まさし）

1933年	福島県に生まれる.
1956年3月	早稲田大学第一理工学部電気工学科卒業.
1956年4月	横河電気製作所株式会社（現 横河電機株式会社）入社. 主として輸出業務に従事.
1973年	英検1級取得.
1981年	同社退職.
1982年	翻訳会社「有限会社海外テクノサービス」設立.
1986年	翻訳会社「株式会社ユーテクノサービス」を横河電機株式会社と共同出資で設立.
1992年	同社退職.
1993年	う沼翻訳事務所設立 現在に至る.

知りたいことがすぐわかる
数・式・記号の英語

　　　　　　　　　平成15年12月30日　　発　　　行
　　　　　　　　　平成31年 3月20日　　第9刷発行

著作者　　鵜　沼　　　仁

発行者　　池　田　和　博

発行所　　丸善出版株式会社

〒101-0051　東京都千代田区神田神保町二丁目17番
編集：電話(03)3512-3264／FAX(03)3512-3272
営業：電話(03)3512-3256／FAX(03)3512-3270
https://www.maruzen-publishing.co.jp

© Masashi Unuma, 2003

組版印刷・製本／藤原印刷株式会社

ISBN 978-4-621-07340-7 C2082　　　　　Printed in Japan

JCOPY 〈(一社)出版者著作権管理機構　委託出版物〉

本書の無断複写は著作権法上での例外を除き禁じられています．複写される場合は，そのつど事前に，(一社)出版者著作権管理機構（電話 03-5244-5088, FAX 03-5244-5089, e-mail：info@jcopy.or.jp）の許諾を得てください．